ELEMENTS OF
EINSTEIN'S RELATIVITY
THE SPECIAL THEORY

Swapan Kr. Barman

Former Associate Professor and Head, Dept of Mathematics

Maulana Azad College, Kolkata, India.

BlueRose
Publishers
New Delhi • London

First Published in December 2021

ISBN: 978-93-5472-097-0

BLUEROSE PUBLISHERS
www.bluerosepublishers.com
info@bluerosepublishers.com
+91 8882 898 898

Cover Design:
Muskan Sachdeva

Typographic Design:
Ilma Mirza

Distributed by: Blue Rose, Amazon, Flipkart, Shopclues

Dedicated in memory of my parents

Late Sri U. R. Burman

and

Late Smt. H. P. Burman

Front Cover Photograph:

> Albert Einstein in 1905 at his desk in the
> Swiss Patent Office, Berne, Switzerland.

In the year 1905, at his age 26, Einstein published four outstanding papers, viz.,

(i) The Theory of the Photoelectric Effect,
(ii) The Theory of the Brownian Movement,
(iii) On the Electrodynamics of Moving Bodies (The Special Theory of Relativity),
and, as a follow up of the third one,
(iv) Does the Inertia of a Body depend upon its Energy Content? (the paper deriving the famous Mass-Energy relationship: $E = mc^2$).

Einstein received the 1921 Nobel Prize in physics *"for his services to theoretical physics, and especially for his discovery of Law of the Photoelectric Effect."*
The Theory of Relativity was till then a controversial subject.

Revolutionary results of Relativity Theory:

$$\Delta t = \Delta t_0 / \sqrt{1 - v^2/c^2}$$

$$l = l_0 \sqrt{1 - v^2/c^2}$$

$$m = m_0 / \sqrt{1 - v^2/c^2}$$

$$E = mc^2$$

Preface

This book is intended primarily for students in physical science and mathematics at the Undergraduate level. The Special Theory of Relativity is one of the revolutionary theories of the 20th century physics and is one of the founding pillars of the edifice of modern physics. I have endeavoured to make clear, as much as possible, the conceptual difficulties which students usually face in their first learning of the subject. In the course of writing this book, I always kept in mind the problems we had faced, when we, as pupils in Applied Mathematics in the University of Calcutta, had the opportunity of learning the Special Theory of Relativity.

A little bit of the historical development of the theory and contributions of a number of scientists in this course have been touched upon. General readers with a background and curiosity in science may also find the book interesting, provided they set aside the mathematical complexities, if they like to.

I take this opportunity to express my deep regards to my University Professors from whom I had learnt this subject. I am also deeply indebted to my former colleague Prof. Gaur Gopal Datta at Jalpaiguri Govt. Engineering College, Jalpaiguri, West Bengal, who had encouraged me to write the book, and I wish to offer my grateful thanks to him. I hope the book may be of some interest to some readers even more than a century after the advent of this astounding theory.

S. K. Barman
Kolkata, June 2021.

Table of Contents

Chapter 1

Origin of the Theory of Relativity: Special Theory

1.1 Preliminary Concepts and Definitions

The Theory of Relativity originated as a result of efforts to reconcile the laws of Electromagnetism of Faraday-Maxwell with the laws of Mechanics of Galileo-Newton. Its root lay in Electromagnetism and Optics.

Mechanics is mainly concerned with changes in positions of objects in space with respect to time. The positions of an object are measured with reference to something fixed with respect to the person recording the positions, called the **observer.** For instance, a person standing on the ground at a given point may be an observer. Let us consider some examples.

Example 1 — Let us suppose that the aforesaid observer standing on the ground, say O, is standing beside a straight railway track along which a train is moving past him with a uniform velocity, and that there is a second observer, say O', standing beside a window of a carriage of the running train. Suppose the second observer O' simply drops (not throwing) a small piece of stone through the window on the ground. Then, ignoring the air resistance, the observer O' sees that the stone falls on the ground along a straight vertical line. The observer O on the ground, however, sees that the stone falls on the ground along a parabolic path*.

---

*It is known from Particle Dynamics that "The path (trajectory) of a projectile in vacuum is a *parabola* ", and also that "The path described by a particle, projected horizontally in vacuum from a given point at a certain height above the ground, is a *parabola*."

The following question may then arise: "What 'really' is the trajectory traced out by the stone falling on the ground — a straight line or a parabola?"

It is to be noticed that the answer to this question can only be provided by the aforesaid two observers; there is no third observer. For the observer O, at rest on the ground, the answer would be: "The path is *really* (i.e., with respect to, or relative to, the ground) a *parabola*"; while for the observer O', standing on the moving coach, the answer would be: "The path is *really* (i.e., relative to the body of the coach) a *straight line*".

We see that the parabolic path is as real for (relative to) the observer O as the straight line path is for (relative to) the observer O'.

Example 2 — Let us suppose the train, as described in the Ex.1, is running with a uniform velocity v along the track. This velocity v is measured with respect to the ground, that is, relative to the observer O standing on the ground. Suppose the observer O throws a cricket ball horizontally with a velocity u parallel to and towards the direction of motion of the train. Then, the velocity of the ball as seen by (i.e. relative to) the observer O' standing in one of the coaches of the train would be

$$u' = u - v.$$

If O throws the ball opposite to the direction of motion of the train, the velocity of the ball, as measured by O' would be

$$u' = u + v.$$

Example 3 — Suppose now, the observer O' standing in a carriage of the moving train, throws a cricket ball horizontally with a velocity u' along the direction of motion of the train. The velocity u' is measured with respect to the carriage carrying the observer O'. Then the velocity of the ball as noted by (i.e. relative to) the observer O on the ground would be

$$u = u' + v .$$

If the ball is thrown by the observer O' standing in the carriage with the velocity u' opposite to the direction of motion of train, the velocity of the ball relative to the observer O would be

$$u = u' - v .$$

It is clear from the above three illustrations that, observations or measurements can only be made with respect to *some system of reference.* For the observer O the *ground* is the reference system, and for the observer O' the *railway coach* is the reference system

It should further be noted here that, in order to fully describe the *motion of an object* it is necessary to specify its *position* with respect to *time.* At every point of the path traced out by the object (the stone or the ball in the aforesaid examples) we are to determine the *corresponding time* at which the object occupies a *specific position.* For this purpose, we need, in each of the above three examples, two identical and synchronized *clocks* — one clock with the observer O standing on the ground and the other clock with the observer O' standing in the running train.

We now describe what is meant by **a frame, or a system, of reference.** *A frame, or a system, of reference* in space is a set of rectangular Cartesian Coordinate axes OX, OY, OZ together with a clock fixed at the origin O, and an observer being situated *at rest* at the origin.

In order to work with *space and time together* we, instead of considering the positions of an object in space and the corresponding times, separately, combine them together by means of what are known as *events.* An **event** is an occurrence (or happening) at a certain point in space at a certain instant (moment or point) of time. By an occurrence we may mean even only the presence of a particle, or an object, at a particular position in space at a particular moment of time, so that the motion of the particle, or the object, may be described as a *continuous sequence of events.*

Examples of *events* may be the *emission* of a flash of light from some source, or the *absorption* of a light signal by some receiver, or say, a *collision* between two particles at a certain point in space at a certain instant of time. In a frame of reference OXYZ, an *event* is specified by four numbers (x,y,z,t) denoting the space coordinates $x,y,z,$ and the corresponding time of occurrence t of the event in the system.

1.2 Historical Background.

Galileo (1564-1642), by means of his experiments perceived that: In the absence of a force, an object at rest or in uniform motion would continue in that state of rest or of uniform motion, and also that a force is necessary to change that state. The inherent tendency of an object to resist any change in its state of rest, or of uniform motion, was termed as its *property of Inertia* by Kepler (1571-1630).

Newton (1642-1727) refined these ideas and precisely enunciated them as his *First Law of Motion.* This law, also known as the *Law of Inertia,* or the *Fundamental Law of Mechanics of Galileo-Newton,* states that:
"An object free from the action of any forces continues in its state of rest or of uniform motion in straight line."

Now, to apply this law we need some *frame or system of reference* with respect to which the state of rest, or of uniform rectilinear motion, of an object is to be described.

"A *system or frame of reference* in which the Fundamental Law of mechanics of Galileo-Newton (Law of Inertia) *holds,* is said to be a *Galilean system (or frame) of Reference, or an Inertial system(or frame) of Reference."* By definition, in such a system then, an object that is acted upon by no external forces, would either be at rest, or move with a uniform velocity in a straight line.

It consequently follows that: "If a system, or frame, of reference (coordinate frame) is an *inertial system,* then every other reference system (frame), which is *in uniform rectilinear motion* with respect to the former, is also an *inertial one*."

1.3 Galilean Principle of Relativity.

Let us imagine a ship sailing on calm water at a uniform speed along a fixed direction without any turn, and a person carrying out some mechanical experiments (for example, studying the motion of a pendulum, or the paths of falling objects) within the ship's hull, without looking outside the ship. On the basis of mechanical experiments being performed inside the ship, as well as on the shore, Galileo arrived at the following conclusion:

"All mechanical experiments performed within a ship moving with a uniform velocity (unaltered in speed and direction) relative to the shore, would yield precisely the *same results* as given by similar experiments performed on the shore." This means that, "One observer on the shore and the other observer inside the ship, would not be able to ascertain *whether the ship was moving relative to the shore*, simply by comparing the results of the same set of experiments carried out on the shore, as well as, inside the uniformly moving ship."

Generalizing these observations Galileo enunciated his **Principle of Relativity** as follows: "Any two observers moving at a uniform speed in the same direction (i.e. with uniform velocity) *relative to each other*, would obtain the *same set of results* for all mechanical experiments performed by them." (it being obviously understood that the two sets of experimental apparatus, used by the two observers, separately moved along with them.)

In a more scientific language this Principle may be stated as:
"All the laws of *mechanics* (laws governing mechanics) assume the *same form* in all frames of reference which are *inertial ones*"; that is, "All inertial frames are *equivalent* from the point of view of *mechanics.*"

In other words, the Principle is: "In Classical mechanics (mechanics of Galileo-Newton) all inertial frames are *equivalent.*"
This Principle (formulated by Galileo in the 17th century) is *one of the building blocks* of Einstein's Theory of Relativity.

1.4 Classical Mechanics and the Galilean Transformation

So far, we have been concerned with the *First Law of Motion* of Galileo-Newton. We now consider the *Second Law of Motion* of Newton. This law states that:

"The rate of change of momentum of a particle, (or of an object), is proportional to the applied force and takes place in the direction of action of the force."

Let us consider two *inertial systems* of reference K(Oxyz) and K'(O'x'y'z') such that the x and x'-axes are coincident, the axes of y, y' are parallel and so are the axes of z, z'. Let the system K' be moving with a *uniform velocity* v relative to the system K along the common x-x'-axis. Further, let two identical clocks be fixed at the respective origins of the two systems, and let the clocks be so synchronized that the origins O and O' coincide with each other at the time $t = t'= 0$. Such frames of reference are said to be in **standard configuration** (Fig.1).

To consider the Second Law of Newton, we need the transformation relations between the coordinates of any given *event*, say (x,y,z,t) in K and (x',y',z',t') in K'. As the frame K' is moving with a uniform velocity v relative to the frame K, the required transformation relations, at any time t, are the following:

$$x' = x - vt, \quad y' = y, \quad z' = z, \quad t' = t. \quad \ldots\ldots \quad (1.1)$$

As *time* is regarded as *absolute and universal* in the Newtonian theory, the last of the Eqs.(1.1)is taken to be self-evident. However, it may be useful to assign different symbols t and t' for measuring the times in the two systems K and K', as will be found later. The set of relations in the Eqs.(1.1), giving the transformation relations from (x, y, z, t) to (x', y', z', t'), is known as the *Galilean Transformation*.

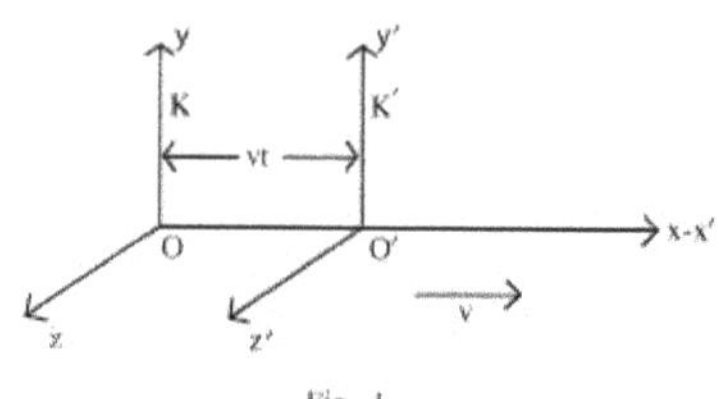

Fig . 1

If we assume the system K' to be at rest, then the system K would be moving with a uniform velocity $-v$ with respect to K', and the *Galilean transformation* relations from the frame K' to the frame K would be

$$x = x' + v\,t', \quad y = y', \quad z = z', \quad t = t'. \quad \ldots\ldots \quad (1.2)$$

Now, let Newton's Second Law be valid in one of the two systems K and K', say in K. Let a force $\vec{F} = (F_x, F_y, F_z)$ be acting on a particle of mass m in the system. Then according to Newton's Second Law

$$F_x = m\frac{d^2x}{dt^2}, \quad F_y = m\frac{d^2y}{dt^2}, \quad F_z = m\frac{d^2z}{dt^2}. \quad \ldots\ldots \quad (1.3)$$

Since, using the Eqs. (1.1), we get

$$\frac{d^2x'}{dt'^2} = \frac{d^2(x - vt)}{dt^2} = \frac{d^2x}{dt^2} \qquad (v \text{ being uniform}),$$

by assigning the same values for the mass and the acting force, i.e., by writing $m' = m$ and $\vec{F'} = \vec{F}$, or $(F_x', F_y', F_z') = (F_x, F_y, F_z)$, we obtain the Eqs(1.3) as

$$F_x' = m'\frac{d^2x'}{dt'^2}, \quad F_y' = m'\frac{d^2y'}{dt'^2}, \quad F_z' = m'\frac{d^2z'}{dt'^2}, \quad \ldots\ldots \quad (1.4)$$

so that the Second Law retains the *same form* in the system K' as well.

Galilean transformation also furnishes the *Classical Law for Addition of Velocities.* Let u and u' be the velocities of a moving particle with respect to the inertial frames K and K' respectively, where v is the velocity of K' relative to K along their common x-x'- axis. Then from the Eqs.(1.1) we have

$$\frac{dx'}{dt'} = \frac{d}{dt}(x - vt) = \frac{dx}{dt} - v$$

or
$$u' = u - v, \quad \ldots\ldots \quad \ldots\ldots \quad \ldots\ldots \quad (1.5)$$

and from the Eqs.(1.2) we have

$$\frac{dx}{dt} = \frac{d}{dt'}(x' + vt') = \frac{dx'}{dt'} + v$$

or $$u = u' + v. \qquad \ldots\ldots \quad \ldots\ldots \quad \ldots\ldots \qquad (1.6)$$

The above two results are in agreement with our common experience, as already illustrated in Ex.2 and Ex.3 given in the Sec.1.1, K being the inertial frame fixed with respect to the ground and K' the frame fixed with the train, running with a velocity v relative to the ground.

Newton's *Third Law of Motion* plays somewhat less significant role in the theory. However, as forces are the same in magnitude and direction in both the systems K and K', the Third Law remains valid in both the systems whenever it is valid in either.

It is thus clear that, if Newtonian mechanics is *valid* in any *one inertial frame*, then it is *valid* in *every such frame*. As two inertial frames can be connected by a *Galilean transformation*, viz., the Eqs. (1.1), or Eqs. (1.2), it then follows that:
"The laws of Classical mechanics (mechanics of Galileo-Newton) remain *invariant* (remain the *same in form*) under a Galilean transformation".
This is the *Principle of Galilean Relativity*, as already stated in the Sec. 1.3, in a little different but equivalent, form.
Summing up, the Principle of Relativity of Classical Mechanics can be stated thus:
"No *mechanical experiment* can distinguish between two inertial frames, all such frames being *equivalent* for the description of *mechanical processes*".
Expressed in a somewhat mathematical language this Principle is the following:
"The *fundamental equations* that determine the sequence of events in any *mechanical process*, remain *invariant* under a Galilean transformation".

1.5 Electromagnetism and the Galilean Transformation

Classical mechanics, founded on the laws of Galileo and Newton, had been very successful, for about three centuries, in accounting for the motion of extra-terrestrial objects like the moon, the planets and comets, as well as in solving all the terrestrial problems of mechanics. However, investigations in the electromagnetic phenomena by scientists in the 19th century, gave rise to the need for a review of the Classical Newtonian ideas.

Based on the experimental results mainly of Faraday (1791-1867) and of other physicists, Maxwell (1831-1879) developed a set of four equations, which brilliantly explained all the electromagnetic problems known up to that time, and further, from these equations Maxwell derived the *electromagnetic wave equation,* which showed that disturbances in the electromagnetic field are propagated through space in the form of waves that carry energy (1856-73). This equation contains a *constant c,* which was identified with the velocity of propagation of the electromagnetic waves in vacuum. Moreover, this velocity *c* agreed very closely with the velocity of light (regarded as waves since the 17th century) in empty space, determined experimentally several years earlier by Fizeau (1819-1896) and Foucault (1819-1868). This indicated the identity of light waves with electromagnetic waves.

Long before, light had been regarded as a form of wave motion by Huygens (1629-1695), who had developed the *Wave theory of Light* in 1670-78. After Maxwell's theoretical prediction of the existence of electromagnetic waves, Hertz (1857-1894) created electromagnetic waves in the laboratory and by means of different experiments demonstrated that *light is a form of electromagnetic wave* (1885-89).

1.5.1 The Ether Concept.

As all types of mechanical wave motions investigated in the past, for instance, surface water waves, or sound waves, required some medium for propagation, it was natural for scientists at that time to presume that the light wave, or electromagnetic wave, also needed a medium for its propagation. Physicists believed in the existence of a tenuous medium throughout the whole universe, which was called the *Aether or Ether.* This ether was supposed to:

(a) be absolutely stationary and pervade the entire space of the universe,
(b) exist even inside material bodies,
(c) be perfectly transparent,
(d) be almost of zero density,
(e) offer no resistance to motion of objects through it,
(f) be elastic like in its capacity to store energy,
(g) be able to transmit energy.

Although the *concept (or hypothesis) of ether* had severe difficulties because of its very unnatural properties, its existence was believed to be necessary solely as a medium for the propagation of light or electromagnetic waves. Since the velocities of mechanical waves, like water surface waves, or sound waves, are measured with respect to their respective mediums of propagation, viz., the stationary water-mass, or the stationary air (or any other liquid or solid elastic mediums), the *velocity c of* light (or electromagnetic wave) was supposed to be with respect to an *all pervading stationary medium — the ether.*

In the late 19th and early 20th centuries, it was extremely difficult for physicists to conceive that electromagnetic disturbances (waves) could propagate by itself without any medium. The *hypothesized medium, the ether,* occupying the universal space, was regarded as absolute, unique and stationary, in which the velocity of light (electromagnetic wave) would everywhere and always be

$$c = 3 \times 10^{8} \text{ m/sec.}$$

An inertial frame K *at rest* in this *stationary ether medium* was called a **preferred or privileged inertial frame.** If a frame K′ is moving with a uniform velocity v with respect to this preferred (stationary) frame K, the velocity of light relative to the frame K′ should be other than c. By means of the Galilean (Classical) Velocity transformation relations, viz., the Eqs.(1.5) & Eqs.(1.6), this velocity, say c', would be given by

$$c' = c - v \quad \text{or} \quad c + v,$$

according as the velocity v of K′ relative to K is in the same or opposite direction as that of light c.

Maxwell's equations, the electromagnetic wave equation in particular, are not preserved in form under a Galilean transformation. The wave equation

$$\frac{1}{c^2}\frac{\partial^2 \varphi}{\partial t^2} = \nabla^2 \varphi = \frac{\partial^2 \varphi}{\partial x^2} + \frac{\partial^2 \varphi}{\partial y^2} + \frac{\partial^2 \varphi}{\partial z^2} \qquad \ldots\ldots \qquad (1.7)$$

is transformed into the form:

$$\frac{1}{(c^2-v^2)}\frac{\partial^2 \varphi}{\partial t'^2} = \frac{\partial^2 \varphi}{\partial x'^2} + \frac{\partial^2 \varphi}{\partial y'^2} + \frac{\partial^2 \varphi}{\partial z'^2} \qquad \ldots\ldots \qquad (1.8)$$

so that the velocity c' of light, as measured in the frame K', is given by

$$c'^2 = c^2 - v^2 = c^2(1 - v^2/c^2),$$

i. e., by $\qquad\qquad c' = c\sqrt{1 - v^2/c^2}. \qquad \ldots.. \qquad (1.9)$

Thus, the Laws of Electrodynamics *do not remain invariant* under a Galilean transformation, although the Laws of Classical Mechanics do.

1.6 Selection of the Correct Alternative.

At this juncture, physicists were required to choose one (the correct one) of the following three alternatives:

(i) The Principle of Galilean relativity applies to Classical mechanics but not to Electrodynamics. A preferred (privileged) inertial frame exists for electrodynamics, viz., the unique ether frame at absolute rest.

(ii) The Principle of Galilean relativity applies to both Classical mechanics and Electrodynamics, but electrodynamics, as given by Maxwell's equations, needs modification.

(iii) A Principle of relativity, other than Galilean, exists for both Classical Newtonian mechanics and Maxwell's electrodynamics, but the laws of Newtonian mechanics are in need of modification.

The correct selection from among the above three possibilities could only be done on the basis of experimental results in support of the choice to be made. The experiments could then be divided into the following three categories:

(a) To detect a privileged inertial frame — the stationary ether frame — for the laws of Maxwell's electrodynamics.

(b) To detect deviations from the laws of Maxwell's electrodynamics.

(c) To find deviations from the laws of Classical Newtonian mechanics.

1.7 Search (unsuccessful) for the Absolute Ether.

The first alternative was chosen by most scientists at that time. Various experiments had been carried out to test the motion of the Earth relative to the *absolute stationary ether (hypothesized)*. The attempts however had all failed. The most famous of those experiments was the one performed by Michelson in 1881, and later with improved apparatus and greater precision by Michelson and Morley in 1887.

1.7.1 The Michelson-Morley Experiment (outcomes only).

According to Classical velocity addition rule, if c is the velocity of light in the stationary ether, then an observer moving with a velocity v with respect to this stationary ether would measure the velocity of light as

$$c - v \quad \text{or} \quad c + v,$$

according as the velocity v and the velocity c are in the same or opposite directions.

Maxwell, in 1878, had suggested an experiment, applying the phenomenon of *interference of light*, to detect the motion of the Earth with respect to the stationary ether. However, the precision level of the measurements required for the experiment, was supposed to be unattainable by Maxwell.

But three years later, in1881, Michelson devised an optical instrument, called the *interferometer*, which was capable of detecting the motion of the apparatus (positioned on the Earth) relative to the stationary ether. The experimental set up was as sensitive as to measure a velocity v of the Earth of even 10 km/sec with respect to the stationary ether, although the Earth's orbital velocity around the Sun is about 30 km/sec. The experiment nevertheless failed to detect any such motion.

The *null result* (*nil velocity* of the Earth relative to the stationary ether) of the Michelson-Morley experiment was a great blow to the concept of an absolute ether frame. Attempts were therefore made to preserve the hypothesis of the absolute stationary ether as well as to provide explanation for the null result of the Michelson-Morley experiment. One of these attempts gave rise to the *Contraction hypothesis.*

1.7.2 Fitzgerald-Lorentz Contraction Hypothesis.

The Contraction Hypothesis put forward by Fitzgerald in 1892 and later extended by Lorentz states that:
"Any object moving through the stationary ether with a velocity v is contracted in the direction of its motion by a factor $\sqrt{1 - v^2/c^2}$."

Although very small, the amount of contraction (according to this hypothesis) in the *length of the two arms of the interferometer apparatus* of the Michelson–Morley experiment, which moved (together with the Earth) with a velocity v through the stationary ether, succeeded in accounting for the null result of the experiment. The Ether hypothesis was thus *apparently 'saved'* by the Fitzgerald-Lorentz Contraction hypothesis.

However, an arduous experiment was carried out by Kennedy and later by Kennedy and Thorndike in 1932 with some modifications of the Michelson–Morley experiment. They showed that the null result of the Michelson–Morley experiment had been obtained because of the *equal length of the two arms* of the interferometer instrument, even when the Fitzgerald-Lorentz contraction was taken into account, and it was expected that this null result might not recur with *unequal lengths* of the two arms.

They, therefore, took the two arms of the optical interferometer as much unequal as feasible for the experiment, instead of the two equally long arms in the original Michelson-Morley experiment. Although both the stationary ether and the Fitzgerald-Lorentz contraction were taken into consideration, the Kennedy-Thorndike experiment still produced a *null result* to the order of the experimental error, which corresponded to the velocity of the Earth of only about 10 km/sec relative to the stationary ether.

1.7.3 Ether-Drag Hypothesis

Another effort to retain the ether concept, as well as to account for the *null result*, led to the Ether-Drag Hypothesis. Michelson attempted to account for the null result of his experiment by proposing that the moving Earth dragged the ether adjacent to its surface along with it. It was suggested that:

"The moving Earth or any moving ponderous object drags the adjacent ether along with it."

This hypothesis would automatically yield a null result for the Michelson-Morley and Kennedy-Thorndike experiments. But the Ether-Drag hypothesis came in contradiction with two phenomena, which were known long before, viz. the "Aberration of stars" and "Fresnel-Fizeau Partial-drag of light". These are described in short in the following section.

1.7.4 Evidence against the Ether-Drag Hypothesis.

(a) Aberration of Stars (Stellar Aberration).

This effect was discovered by the astronomer Bradley (1693-1762) as early as in 1728. Due to the orbital motion of the Earth, with a velocity v (say) around the Sun, in order to observe a star directly overhead through a telescope, the telescope tube needs to be inclined at an angle α with the vertical, given by

$$\tan \alpha = v/c = \frac{30 \times 10^3 \text{m/sec}}{3 \times 10^8 \text{m/sec}} = 10^{-4}, \quad \ldots \ldots \quad (1.10)$$

so that $\alpha = 20.5"$ of arc, where c is the velocity of light in the stationary ether. The angle α is known as the *angle of aberration*, and the effect is the same for all stars. Because of this effect, a star fixed in the sky, at a great distance from the Earth, is observed to describe an *elliptic orbit* around a fixed centre in the sky, in the course of one year (the Earth's annual orbit around the Sun being an *ellipse*).

If the ether were dragged by the Earth, then light rays from stars would have been swept away along with the ether adhered to the Earth's surface, and the telescope tube would not have to be tilted from the vertical; that is, there would have been no aberration at all.

(b) Fresnel-Fizeau Partial-Drag of Light.

The second contradiction arises in the case of propagation of light or electromagnetic wave in a moving medium. Fresnel (1788-1827) had predicted in 1817 that *light would be partially dragged by a moving medium*, and derived theoretically a formula for the velocity c' of light in the moving medium. The medium moves with respect to the laboratory, and the velocity c' is relative to an observer at rest in the laboratory. The formula is the following:

$$c' = \frac{c}{n} \pm v_m \left(1 - \frac{1}{n^2}\right) \qquad ..... \qquad (1.11)$$

where c is the velocity of light in the stationary ether, n the refractive index and v_m the velocity of the moving medium. The factor

$$(1 - 1/n^2)$$

is known as the *Fresnel-drag coefficient.* Fizeau in 1851 and 1853, verified this formula in his experiments with moving fluid (water). If the ether were dragged (according to the Ether-drag hypothesis) by the moving water, then light, together with the ether existing in the moving water, would have been dragged (fully) by the moving water, and the velocity of light would have been given by

$$c' = c/n \pm v_m . \qquad ...... \qquad (1.12)$$

This result is in disagreement with Fresnel's formula, Eq. (1.11), which was experimentally verified by Fizeau.

Thus, the Ether-drag hypothesis came *in contradiction* with two well established observed phenomena — the Stellar Aberration and the Partial-drag of Light, and the hypothesis, therefore, was *not acceptable.*

All these considerations about the Ether hypothesis — either an absolutely stationary ether (Preferred Frame hypothesis), or a local ether attached to ponderous objects (Ether-Drag hypothesis) —rendered the Ether Hypothesis *untenable.*

1.8 Modification attempts of Maxwell's Electrodynamics.

At this stage, several efforts were made to modify the laws of Maxwell's electrodynamics, in particular the *Law of propagation of light*, so as to avoid the difficulties or controversies associated with the Ether theories.

1.8.1 Emission Theories.

The theories which sought to modify Electrodynamics are all known as Emission Theories. The most remarkable among them was the one presented by Ritz in 1908. He modified two of Maxwell's equations in such a way as to make the velocity (speed) of light equal to c, only when measured with respect to the light-source. Although the three Emission theories presented, viz., the *Ritz theory* (also called the *Original Source theory*), the *Ballistic theory* and the *New Source theory*, differed among themselves in their predictions, they all proposed that:
"The velocity (speed) of light depends *only on the source* and not on the medium of transmission (the ether), and that this velocity is c relative to the source of emission of light".

This accounted for the null result of the Michelson-Morley experiment. The Ritz theory was also in agreement with the observation of Stellar Aberration and the results of Fizeau's experiment.

However, all the three Emission theories were *contradicted* by two experiments with *extra-terrestrial light sources*, viz., de Sitter's observations with light from binary (double) stars, and the Michelson-Morley experiment performed with star or sun light.

1.8.2 Evidence against the Emission Theories.

(a) De Sitter's Observation of Binary Stars.

A *binary (double) star* is a system of a *pair of stars*, close to each other, revolving in orbits around their common centre of mass. Due to rotation around their common centre, when one member of the doublet moves towards us (the Earth) during half of its period of rotation, the other member moves away from us; and the cycle is reversed during the other half of the motion.

If the velocity of light additively depends on the velocity of the light-source, then the velocity (speed) of light towards the Earth from the approaching star of the pair would be greater than that from the receding star. De Sitter, in 1913, showed by calculation that as a result of this, a *fictitious eccentricity* would appear in the orbits (which are in fact *nearly circular*) of the stars of the doublet. Actually, nothing of this sort has ever been observed in the motion of binary stars.

(b) Michelson-Morley experiment with Extraterrestrial Light Source.

The Michelson-Morley experiment with an extraterrestrial source of light was carried out by Tomaschek in 1924 (using starlight) and by Miller in 1925 (using sunlight). If the source velocity (due to rotational and translational motions of the sun and the star) relative to the interferometer apparatus of the experiment, had any effect on the velocity of light, the Ritz theory would produce *complicated interference fringe patterns*. No such effects, however, were found in either of the two experiments.

From the observations of binary stars by de Sitter and from the results of the experiments of Tomaschek and of Miller, it can then be concluded that: *"The velocity of light is not affected by the motion of the source emitting the light."*

1.9 Evidence of Departure from Newtonian Mechanics.

The velocities of objects that we come across in our everyday life and in the macroscopic world, in general, are all *much less* than the velocity c of light (electromagnetic wave) in empty space. Newton's Laws of Motion hold exactly in the realm of such velocities.

Towards the end of the 19th century, the first microscopic particle, viz. *electron*, was discovered by Thomson in 1894-96 and natural radioactivity discovered by Becquerel in 1896. The *beta ray or beta particles (electrons)* observed among radiations emitted by radioactive substances were found to have *velocities near the speed of light c.*

Kaufmann in 1902, investigated the motion of such *high velocity electrons* in electric and magnetic fields, and found that the motion of such particles, as described by Newton's Second Law of Motion, does not agree with the experimental results. This was the **first instance of departure from the Newtonian mechanics,** and it was observed for particles with *velocities close to that of light.*

1.10 Search for a New Transformation.

The various kinds of experiments described (only in broad outlines) so far, lead us to conclude the following:
(a)The Ether Hypothesis is not tenable.
(b)Modifications of Maxwell's Electrodynamics are *unacceptable.*

We are then left with the only option, viz. selecting the *alternative (iii)* in Sec.1.6, which, therefore, is stated again:
"A Principle of Relativity, other than Galilean, exists for both Classical mechanics and Maxwell's electrodynamics, and the Laws of mechanics of Galileo-Newton need modification."

Thus, by the end of the 19th and in the beginning of the 20th centuries, it turned out to be necessary to modify the Galilean transformation and Newtonian mechanics. Physicists then, looked for a transformation that would *ensure* the *invariance* of *Maxwell's equations.*

Great theoretical contributions in electromagnetic and optical phenomena were made by Lorentz in 1895-1904. Lorentz, who was always in favour of the Ether theory, by using a number of ad hoc hypotheses, set up a transformation under which the Maxwell equations, when transformed from the *stationary ether to a moving inertial frame*, would remain *invariant.*
Later, Larmor also developed similar transformation equations in 1897-1900, and named them after Lorentz. Very similar equations were, however, used by Voigt far earlier in 1887, for his investigations of Doppler's Principle in Wave Motion.

Poincare elaborated and perfected the investigations of Lorentz in 1905-1906.

Lorentz's equations of transformation, from the stationary ether frame K(x,y,z,t) to an inertial frame K'(x',y',z',t') moving with a uniform velocity v relative to the frame K along the common x-x'-axis, are given by

$$x' = (x - v\,t)/\sqrt{1 - v^2/c^2},$$
$$y' = y, \quad z' = z, \qquad \qquad \text{.....} \quad (1.13)$$
$$t' = (t - v\,x/c^2)/\sqrt{1 - v^2/c^2}.$$

Lorentz had always regarded Newton's "absolute time" t as the "true time", and did not consider the variable t' as time in the same sense as the variable t. He called t' the "local time" and considered it to be merely a supplementary quantity introduced as a mathematical artifice.

1.11 Advent of Einstein's Relativity: Special Theory.

The way to the Theory of Relativity (Special Theory) was thus paved in the beginning of the 20th century, and the stage was almost set for the advent of the theory in the year 1905. The theory has been called the *"Special Theory of Relativity"* in order to distinguish it from the generalized theory, known as the *"General Theory of Relativity,"* which came into being about 10 years later in 1915-16.

The final and decisive solution was, however, provided by Einstein in 1905 in his outstanding paper entitled *"On the Electrodynamics of Moving Bodies."* As already seen, a number of experiments, which we have considered so far, were in fact performed several years after 1905. Also, apparently, Einstein was unaware of the related later works of Lorentz and Poincare. He had heard about the result of the Michelson-Morley experiment but not the details. He arrived at the Lorentz transformation through a *radically different course.*

Einstein's paper of 1905 contained all the essential results of Lorentz, Larmor and Poincare, and presented an entirely novel, much more profound and far reaching exposition of the problem. He put forward *two postulates*, from which and by means of his novel analysis of the concepts of *space and time*, he derived the Lorentz transformation and the entire Special Theory of Relativity. In order to have an idea of Einstein's approach we may proceed as follows.

1.11.1 Einstein's Approach.

Einstein looked for a Principle of Relativity that would cover not only the Classical mechanics (as in the Principle of Relativity of Galileo-Newton) but also Electrodynamics, Optics, and indeed all physical processes. This gave rise to the *First postulate.*

Next, it is to be noted that the *constant c,* occurring in Maxwell's equations, refers to the velocity (speed) of propagation of electromagnetic waves in free space (vacuum), and that the electromagnetic wave equation for empty space *does not* contain *any term* related to the motion of any electric or magnetic source of the field, *nor any term* related to the motion of any observer. The question may then naturally arise is:
"What is the reference body, or *reference frame*, the *velocity c* is measured with respect to? "

Physicists in the 19th and early 20th centuries had assumed that a medium — *luminiferous* (meaning *light-transmitting*) *ether* — was necessary for the transmission of light (or electromagnetic waves), and that the *velocity c* of light was with respect to this *ether medium* which was supposed to be at *absolute rest.*

Einstein, on the contrary, considered the *ether* (with its very unusual properties) to be non-existent and redundant as a medium for the transmission of light or electromagnetic waves. According to him, light waves do not require any medium for propagation; electromagnetic disturbances (waves) can proceed through empty space (vacuum) without any supporting entity. The hypothesis of *"luminiferous ether"* was as such *abandoned* by Einstein.

As regards the *reference frame* for measuring the *velocity c* of the electromagnetic waves, Einstein proposed that this velocity should be the *same constant c* relative to *any or every,* inertial frame of reference, whatever. This led to the *Second postulate.*

1.11.2 Postulates of Einstein.

The two celebrated postulates of the Special Theory of Relativity of Einstein are the following.

Postulate 1. The laws of *physics* are the *same* in all inertial frames of reference; there is no privileged inertial system (*The Principle of Relativity*).

This postulate implies that there exists no such thing as "absolute rest", nor any "absolute motion", and that only a ***"relative motion"*** of one object with respect to another has got a physical significance. It is for this reason that Einstein's theory is commonly known as *The Theory of Relativity* or, more precisely, as *The Theory of Relativity of Motion.*

The Ether hypothesis of the 19th century conceived of a privileged frame of reference for all electromagnetic phenomena, viz. the so called *"Ether frame at absolute rest".* The First postulate of Einstein thus *dispensed with* the *"concept of Ether".*

It also follows from the First postulate that:
"It is *impossible* to detect the motion of one inertial system with respect to another by carrying out any *physical experiment* in each of them and then comparing the results of the two".

In short: "All inertial frames are *equivalent* for the description of a *physical process.*"

This is because the experimental phenomena in two inertial systems are governed by the *same laws of physics.*

Postulate 2. The velocity (actually *speed*) of light in empty space has the *same value c* in all inertial systems of reference, irrespective of the *relative motion* between the source of light and the observer (*The Constancy of the Speed of Light*).

The Second postulate appears to be a rather strange proposition. Light from a source, *at rest* with respect to an observer, is known to propagate in free space (vacuum) with a speed

$$c = 3 \times 10^8 \text{ m/sec.}$$

If the light source moves *towards or away* from the observer, then according to Classical mechanics, the speed of light as measured by the observer, would, respectively, be *greater or less* than c, and similarly, if the observer moves towards or away from the source, the speed of light measured by him would, respectively, be greater or less than c.

But according to the *Second postulate*, the speed of light, as measured by the observer in *either case*, would have the *same value c*.

This is contrary to our common sense (the Classical Velocity Addition rule) and difficult to conceive. A number of theories were then presented to account for the experimental facts, known up to that time, without this Second postulate. The most prominent among them was Ritz's Emission theory proposed in 1908-11. However, de Sitter's observation of light from *binary stars*, in 1913, proved conclusively that:

"The speed of light in empty space is independent of the motion of its source".

Ritz's theory and similar other theories were, therefore, subsequently *all rejected.*

1.11.3 Discussion

When Einstein's paper was published in 1905, there was no direct experimental evidence in support of the Second postulate. The basis of this proposition was the *consequence* of Maxwell's Electromagnetic wave equation that: *"The velocity (speed) of light in empty space is the constant c* ($=3 \times 10^8$ m/sec)".

The latter part of the Second postulate, viz., "The speed of light has the *same value c*, irrespective of the motion of the observer", follows from the First postulate because, it *matters the same,* whether the light source (double star) proceeds toward or recedes from the observer (on Earth), or the observer proceeds toward or recedes from the source; it is the *relative velocity* between the light source and the observer, that *only, indeed matters.*

Some physicists have the opinion that the Second postulate is incorporated in the First one, as Maxwell's laws, being the fundamental laws of electrodynamics, are *laws of physics* and they lead to a *constant velocity of light c,* in free space.

As to the common sense idea, it can be said that, the speeds which we face in our day to day life (wherein our common senses grow and develop) are *much smaller* than the speed of light c, and therefore, it should be the good scientific sense that: "The common sense rule (Classical rule) for the Addition of two velocities which we are accustomed to, *might not* be the *same* when, among the two velocities to be added, one is the *velocity of light c".*

Measuring the speed of electromagnetic radiation emitted from the annihilation of rapidly moving positrons (antielectrons), in 1963, and also measuring the speed of gamma radiation (electromagnetic radiation) emitted from the decay of high speed pie-zero mesons (speed $= 0.99975\,c$) in 1964, it has been decisively established that:
"The speed of light (electromagnetic radiation) is, in fact, independent of the motion of the source".

Chapter 2

Time, Simultaneity and Synchronism

In the pre-relativistic days, *time* was believed to be a continuum within which the entire universe exists and its various changes take place. It was regarded as absolute (unchangeable), universal (same everywhere) and independent of any changes in the physical world. In his *"Philosophiae Naturalis Principia Mathematica"*(1687), Newton wrote: *"Absolute, true and mathematical time, of itself, and from its own nature, flows equably without relation to anything external".*

In his 1905 paper, Einstein pointed out that it was our **concept of time,** prevailing till then, that was the *key factor,* and *needed revision.* He had worked for 10 years before 1905, and after many futile attempts, he later said, *"..... at last it came to me that time was the suspect."*

2.1 Time and its Measurement.

Any change in the physical universe can only take place through the *passage (flow) of time,* and the passage of time can only be detected and measured by means of a *change in a physical process* in the universe. For instance, the rising and setting of the sun take place through the time interval, on the average, of about half a day; and correspondingly, the time period of about half a day may be measured by the physical phenomena of rising and setting of the sun.

To describe more accurately, the shadow on a sundial moves with the passage of time, and correspondingly, the flow of time can be measured by the displacement of the shadow on the dial.

Physics is concerned with recording the *instant (point) of time* of an event, or measuring the *interval(duration) of time* of a physical process, rather than with any non-physical (meta-physical or philosophical) notions of time. Mechanical watches or clocks are widely used in time measurements, and by the *time of an event* we mean the *reading* (position of the pointers or hands) *of a clock* situated in the immediate proximity of the place of occurrence of the event. In any process of recording or measuring the time, what we tacitly employ, is the concept of *simultaneity.* This is elaborated in the following section.

2.2 Simultaneity of Events.

Let us consider the motion of a train running from a station A to a distant station B. The statement: "The train leaves the station A at 7:00 a.m." means that "The departure of the train from the station A and pointing of the small hand of the station clock at the station A exactly toward 7:00", are *simultaneous* events. Similarly, "The train reaches the station B at 7:10 a.m." means that "The arrival of the train at the station B and the positions of the hands of the station clock at the station B indicating 7:10", are *simultaneous* events.

Each of the above two examples makes clear what we understand by "*Simultaneity of events* occurring at a *point* in space". In the first case, the point of occurrence is the point A, and in the second, it is the point B. The two illustrations also show how we apply the concept of *simultaneity of an event* and the *pointing of the hands of a clock* in the immediate vicinity of the *event,* for reading the *time of an event* occurring at a *single point.*

Now, we are to make ourselves clear about the meaning of "*Simultaneity of events* occurring in a frame of reference, i.e., at *different (distant) points* in the frame". We proceed as follows.

Let two posts of signal lights be situated at the points R and G, far apart from each other, along a straight railway track on the ground. Two lights R_1 (say red) and G_1 (say green) are fixed on the top of the posts at R and G respectively, at equal heights over the ground. Let an observer O be standing on the ground at the mid-point M of the line segment RG. The observer O is provided with an instrument (for example, two mirrors inclined at right angle to each other and each at an angle 45^0 to the line RG) which enables him to observe the two lights R_1 and G_1 together. Let the two signal lights R_1 and G_1 on the posts at R and G be flashed. If the observer O at the mid-point M of the segment RG receives the two flashes at the *same time*, that is, if the two beams of light from R_1 and G_1 meet each other at M(O), then the events of flashes at R_1 and G_1 are defined to be *simultaneous events* with respect to the ground — the system of reference (Fig.2).

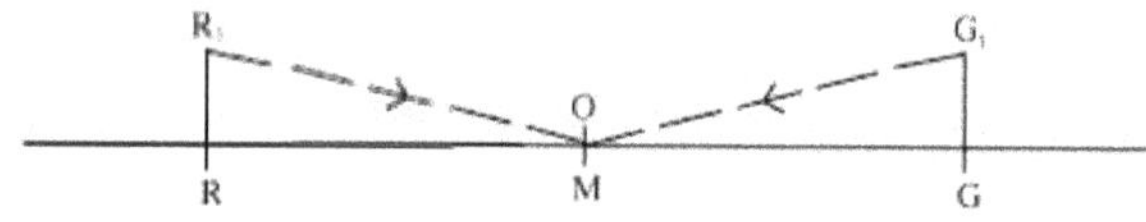

Fig . 2

2.3 Synchronism of Clocks.

We again consider the preceding example of the train leaving the station A at 7:00 a.m. and reaching the station B at 7:10 a.m. We assume that the two station clocks at A and B are of identical make, so that they run at the *same rate*. Suppose we want to determine the time (interval) of travel of the train from the station A to the station B. We, however, may not be correct if we conclude that this time interval is 10 min., unless we ensure that the two station clocks at A and at B are *synchronous*, that is, they *indicate the same time at any particular instant.* For, if the clock at B is slower by one min. than the clock at A, we shall count the time of travel of the train as 9 min., and if the clock at B is faster by one min. than the clock at A, the time of travel will be counted as 11 min.

It is then necessary to have the *two clocks*, situated at distant places A and B, *synchronized* with each other. This can be done in the following way.

Let l be the distance between A and B, and c the *constant speed* of light (electromagnetic wave) in empty space. Let an observer at the station A read a *time t_0* in the clock A and instantly (in zero time) send it by means of light signal (electromagnetic wave) to the observer at the station B, and who, on receiving the signal, immediately (in zero time) sets the clock B to the *time $t_0 + l/c$,* the time taken by the light signal to reach from A to B being l/c. The *two clocks* at the stations A and B are now *synchronized.*

In a similar manner, the clock B can be synchronized with another remote clock C (of identical make as the clocks B and A), and the clock C also, therefore, gets synchronized with the clock A, and so on. We can thus have *identical synchronized clocks* placed at any number of points in a frame of reference (here the ground). Any one of these clocks (all *synchronized* with one another) would then show the time in its immediate proximity, as well as at any distant place in the reference frame.

It should now be noted that nothing, as yet, has been said about *clocks in the running train* (we assume that there are clocks in the train) and about the times recorded by them.

We suppose that all the clocks in the train have been made *synchronous* among themselves and also with the clock of the station A (by the process of synchronization described above) *before the departure* of the train from the station A. This means that the train departed from the station A at 7:00 a.m. according to the *station clock* at the station A *as well as* according to *any clock in the train.* The following question may then arise.

"What would be the *reading of a clock* in the *running train* (the *moving frame* of reference), when the train arrives at the station B at the time 7:10 a.m., as shown by the *station clock* at the station B? Would the *clocks in the train* also show the *same time* 7:10 a.m., or not?"

This question will be discussed later in the Sec.2.5.

2.4 Relativity of Simultaneity.

It will now be shown that: "Two events which are *simultaneous* in one system of reference, *may not be so* in another system."

We recall the illustration of the signal lights R_1 and G_1 fixed on the posts at R and G respectively, situated along the straight railway track as given in the Sec.2.2. Let the distance RG be sufficiently long, and let the train be running in between R and G from R towards G.

Let the lights at R_1 and G_1 be *flashed simultaneously* with respect to the ground (one system of reference, say K). This, according to the definition of simultaneity (given in Sec.2.2), means that the two beams of light from R_1 and G_1 would meet each other at the *mid-point M* of the line segment RG, that is, would *meet simultaneously* an observer O standing on the ground at M. Let O' be the position of an observer on the running train (the other system of reference, say K'), which coincides with the point M (observer O) on the ground just when the lights at R_1 and G_1 are flashed together with respect to the ground (Fig. 3a).

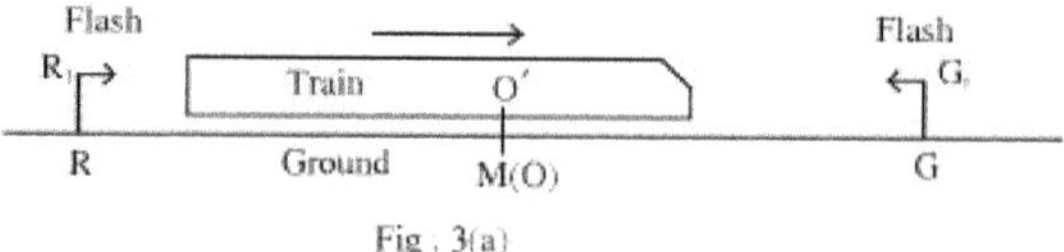

Fig . 3(a)

As the train runs towards the signal post at G and away from the post at R, the observer at O' standing on the train, moves towards the ray of light coming from G_1 and ahead of the ray coming from R_1. Hence, the observer O' would see the ray of light from G_1 before the ray from R_1. The observer O' in the running train would then conclude that the light flash at G_1 took place earlier than that at R_1, that is, the flashes of light occurring at G_1 and R_1, as seen by O', were *not simultaneous* (Fig. 3b).

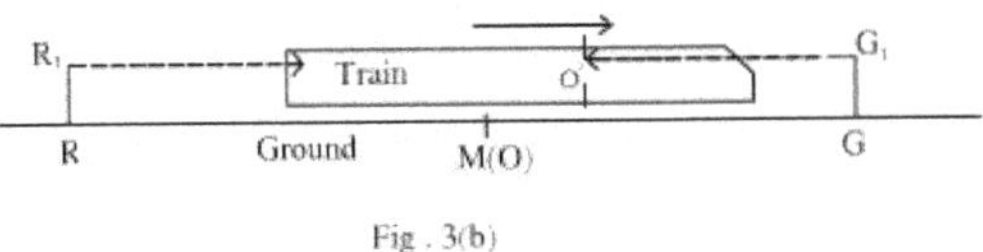

Fig . 3(b)

Next, we suppose that an observer O' is standing at the *mid-point* M of a long carriage R'G' of the train running in the direction from R to G, and that there are two lights fixed at the far ends R' and G' of the carriage. Let the two lights at R' and G' be turned on *together* inside the moving carriage, and let the observer O', standing at the point M of the running carriage, coincide just at that instant, with *some observer* O on the ground. Then the observer O' at M would receive the flashes from R' and G' *simultaneously*. This, by definition, means that the two light beams from the ends R' and G' of the moving carriage would meet each other at the mid-point M(O') of the carriage (Fig.4a).

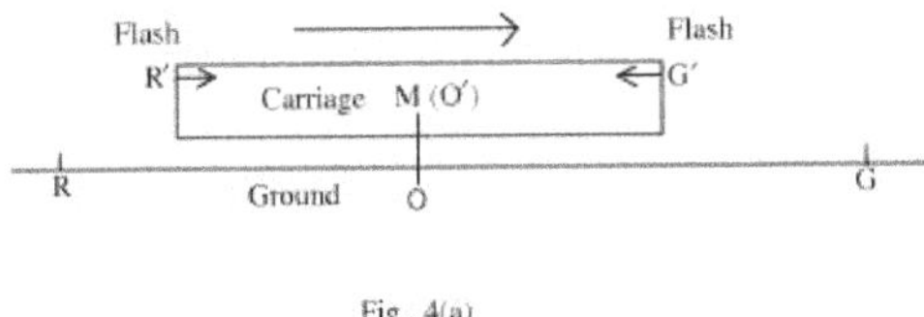

Fig . 4(a)

Here, it is to be borne in mind that, according to Einstein's Second postulate [Sec.1.11], "The velocity (speed) of light in empty space is the same, irrespective of the motion of the light source or of the observer"

Now, as the carriage moves in the direction from R to G, that is, in the direction R' to G', the beam of light from R' moves past the observer O on the ground before reaching the observer O' in the train, and the beam from G' reaches O' before arriving at O. The observer O on the ground would then conclude that the light flash at R' took place earlier than the flash at G', so that the flashes of the lights at R' and G' in the moving carriage were *not simultaneous* as observed by O on the ground (Fig.4b).

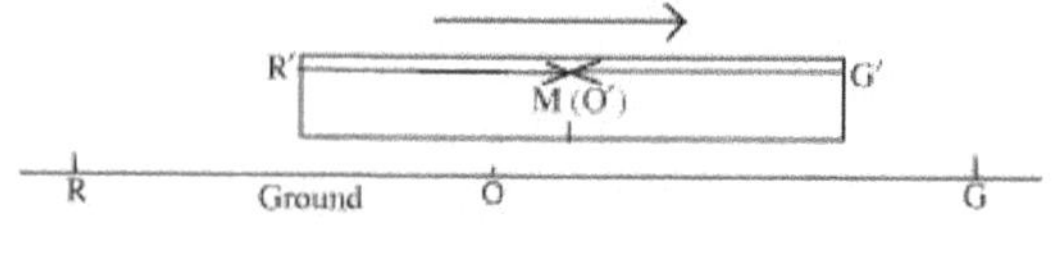

Fig . 4(b)

We thus arrive at the following result:

"Events which are *simultaneous* with respect to one reference frame, say K (here the ground), *may not be simultaneous* relative to another reference frame, say K' (here the running train), and vice-versa."

This result shows that simultaneity of two events is *relative*, and this is called the *Relativity of Simultaneity*.

2.5 Relativity of Time Interval.

We again recall our familiar illustration of the train running along a straight track with a uniform velocity, say v , from the left hand to the right hand side. Let a *light source* and a *clock* be placed at a point S_1 on the floor of a carriage, and let a *mirror* be fitted on the ceiling at a point M_1 vertically above the point S_1. A light signal is transmitted from the source at S_1 toward the mirror at M_1, from which it is reflected back along the line M_1S_1 and reaches again the point S_1. An observer *in the train* sitting at the point S_1 registers this time interval, say Δt_o, by the clock at S_1, and the time taken by the light signal to travel from S_1 to M_1 is therefore *half* of the time interval Δt_o .

Now, because of the motion of the carriage from the left to the right side, the light source S_1 and the mirror M_1 move together to the respective positions S_2 and M_2 toward the right (as observed by *another observer standing on the ground*), when the light signal from S_1 reaches the overhead mirror at M_2. Thereafter, the source and the mirror move together to the respective positions S_3 and M_3, when the light, being reflected from the mirror at M_2, arrives at the source at S_3. The triangle $S_1M_2S_3$ is thus an isosceles triangle with $S_1M_2 = M_2S_3$. The observer on the ground records the time interval, say Δt, for the light to traverse the path $S_1M_2S_3$ by means of *two clocks* at rest on the ground, one situated at the point S_1 and the other at the point S_3, each synchronized with the other.

The time taken by light to travel from S_1 to M_2, as determined by the observer on the ground, is therefore, half of this time interval Δt. Here, S_1S_3 is the distance along the track travelled by the observer in the train carriage with velocity v, as the light ray traverses the path $S_1M_2 + M_2S_3$ (Fig.5).

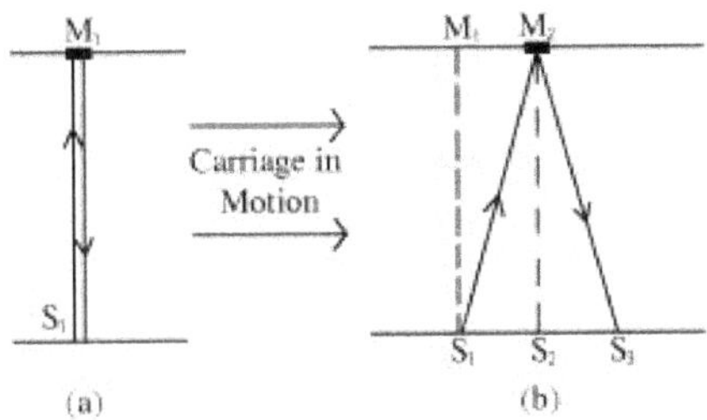

Fig . 5

Fig.5. Path of light ray starting from S_1 and as seen by:
    (a) an observer in the moving carriage,
    (b) an observer on the ground.

We now use Einstein's Second postulate — "The velocity (speed) of light in empty space is the *constant c* in both the frames of reference, viz., the *moving carriage (running train)* and the *ground.*" The path $S_1M_2+M_2S_3$ of light being longer than the path $S_1M_1 + M_1S_1$, we have

$$c.\, \Delta t_0 < c.\, \Delta t, \quad \text{or} \quad \Delta t_0 < \Delta t. \qquad ..... \quad ..... \quad (2.1)$$

We also have

$$S_1M_1 = c.\frac{\Delta t_0}{2}, \quad M_1M_2 = S_1S_2 = v.\frac{\Delta t}{2}, \quad S_1M_2 = c.\frac{\Delta t}{2}.$$

Then from the right angled triangle $\Delta S_1M_1M_2$, it follows that

$$S_1M_1^2 + M_1M_2^2 = S_1M_2^2,$$

or
$$\left(\tfrac{1}{2}c\,\Delta t_0\right)^2 + \left(\tfrac{1}{2}v\,\Delta t\right)^2 = \left(\tfrac{1}{2}c\,\Delta t\right)^2$$

or
$$c^2(\Delta t_0)^2 = (c^2 - v^2)(\Delta t)^2$$

or
$$\Delta t_0 = \Delta t \sqrt{1 - \frac{v^2}{c^2}} < \Delta t, \qquad \dots\dots \quad (2.1a)$$

and also
$$\Delta t = \frac{\Delta t_0}{\sqrt{1 - v^2/c^2}} > \Delta t_0, \qquad \dots\dots \quad (2.1b)$$

The clock at S_1 in the train carriage was *synchronized* with a clock on the ground *before the train started* to move on the track, and all clocks in the moving carriage (as well as in the moving train) are *synchronized* with one another, and also, all clocks on the ground are *synchronized* among themselves.

It is thus seen that for the *event*—"The departure of a ray of light from a given point on the floor of the moving carriage and its subsequent return to the same point, after reflection from a mirror fixed vertically above on the ceiling of the carriage"—the *required time interval,* as measured by the clock in the moving carriage gets *shorter* by a factor $\sqrt{1 - v^2/c^2}$, than the *time interval* for the *same event,* as determined by the *clocks on the ground.*

In other words, for a *given event,* the required *time interval* as determined by the *clocks on the ground* becomes *longer (dilated or expanded)* by a factor $1/\sqrt{1 - v^2/c^2}$ than the *time interval required* for the *same event* as measured by any *clock in the running carriage (train).* That is, the clocks in the running train get *slower* than the clocks on the ground.

This means that in a *moving frame of reference* (here the running train) the *time* passes (flows, elapses, or runs) at a *slower* pace than the *time* in a *stationary frame* (here the ground). This is known as the *Relativity of Time,* or the *"Einstein Dilation of Time (Time Interval)".* In short:

"Any clock runs *fastest* when it is *at rest* with respect to the *observer.* A clock *moving* relative to an observer runs at a *slower rate* than a clock *at rest* relative to the observer ".

Every inertial system of reference, therefore, *contains its own time.* Whenever we make any statement about the *time* of an event, we must necessarily specify the *reference system* to which this *time* refers. This is the *concept of time in physics.*

2.5.1 Proper and Non-Proper Times.

The time (or time interval) recorded by a *single clock at rest* relative to an observer, is called the *"proper time (or time interval)"* for the observer. It is the time interval between *two events* occurring at the *same place*, measured by a *single clock* at that *point.* In the illustration given above, the *time t_o (or time interval Δt_o)* as registered by the clock at the point S_1 in the train is the *proper time* for the *event* taking place at the point S_1.

The time (time interval) Δt between *two events,* occurring at *two different points* S_1 and S_3, and determined by *two different clocks* at those points, is the *non-proper time (time interval)* in respect of the observer on the ground.

Example 1 — We recall our old illustration of a train running from a station A along a straight track up to a distant station B, given in the Sec.2.3. The train leaves the station A at 7:00 a.m., according to the station clock at A, as well as according to a clock in the train, say that in the locomotive. The train reaches the station B at 7:10 a.m. according to the station clock at B. The question was:
"What would be the time according to the clock in the train (locomotive) when the train arrives at the station B?"

We suppose that the train runs with a uniform velocity v from the station A to the station B. Let the time interval for the *event* that the "Train travels from the station A to the station B", be Δt_o, as recorded by the clock in the locomotive, and let it be Δt, as determined by the two clocks (synchronized) situated at the two stations A and B.

Now Δt_o being the time interval measured by a *single* clock attached with the running locomotive, is the *proper time* for the aforesaid *event of travel,* and Δt, being the time interval for the *same event* as determined by *two different* clocks stationary in the ground, is the corresponding *non-proper time*. We therefore have:

$$\Delta t = 10 \text{ min,}$$

and
$$\Delta t_0 = \Delta t \sqrt{1 - v^2/c^2} < \Delta t, \quad [\text{by Eq. (2.1a)}]$$

so that
$$\Delta t_0 < 10 \text{ min.}$$

Let us then, for instance, suppose that $\Delta t_0 = 9$ min. 59 sec. This means that the time shown by the clock situated in the locomotive, when the locomotive arrives at the station B, is 7:09:59 a.m. That is, the clock in the moving train has fallen behind by 1sec. from the clock at the station B in 10 min. We thus have the following result:

"The clock in the moving locomotive, as well as all clocks in the moving train, run at a *slower rate* than the clocks which are stationary on the ground."

Example 2 — In the above illustration, it may be interesting to find the velocity v which the train was running with. Since,

$$\Delta t_0 = \Delta t \sqrt{1 - v^2/c^2}, \quad [\text{by Eq. (2.1a)}]$$

we have
$$1 - v^2/c^2 = (\Delta t_0/\Delta t)^2 = (599/600)^2 = 0.9966,$$
or
$$v^2/c^2 = 0.0034$$
or
$$v/c = \sqrt{0.0034} = 0.0583$$
so that
$$v = 0.0583c = 0.0583 \times 3 \times 10^5 \text{ km/sec}$$
$$= 0.1749 \times 10^5 = 17490$$
$$\approx 17,500 \text{ km/sec.}$$

The distance, say l, of the station B from the station A, as measured by the observers on the ground is then given by

$$l = AB = v.\Delta t = v \times 10 \text{ min}$$
$$= 17,500 \text{ km/sec} \times 10 \times 60 \text{ sec}$$
$$= 10,500,000 \text{ km} = 10.5 \text{ million km.}$$

The extraordinarily large values of the velocity v of the train and the distance l between the stations A and B, show that the example under consideration is that of an "*imaginary thought experiment.*" The results obtained, however, are all fully valid, although hardly perceptible in our ordinary life, as the velocities we deal with in everyday life are far less than the velocity c of light.

2.6 Relativity of Length.

We are used to measuring the *lengths* of objects, or the *distance* between two points, when the said objects, or the points, are *at rest with respect to us.* In such a case, we place a unit measuring stick along the length to be measured the necessary number (including fraction) of times, and from this number we determine the required length.

Now we want to measure the *length* of an object which is *moving with respect to us.* The job is a bit difficult one, and to carry out this we proceed as follows. We again recall the illustration of the train running with a uniform velocity v along a straight track, now parallel to the platform of a station.

2.6.1 Moving Object, Observer at Rest.

At first, we want to determine the length l of a train running with a uniform velocity v, as observed from a fixed point on the ground or the platform.

Let l_0 be the length of the train from its rear end R up to the front end S of the locomotive, as measured by a *passenger* O', in the usual manner, that is, by consecutive laying down of a metre stick on the floor of the train along its length from R to S (or from S to R). In this measuring process, the passenger (or the *observer*) O', is *at rest* relative to the train.

This length l_0 can *alternatively* be determined by the observer O' *sitting* (*at rest*) in the *moving train* as follows (Fig.6).

The observer O' measures the *time required* by the "Entire length of the *running train* to move past *any fixed point,* say F, on the platform (ground)". He determines this time interval by means of *two synchronized clocks* fixed at the end points S and R of the train. Let two persons at S and R (say, the engine driver and the train guard) register the times t_S and t_R as the *clocks* at S and R, respectively, move past the fixed point F. Then,

$$t_R - t_S = \Delta t \text{ (say)}$$

is the *required time interval.* The length l_0 is thus *alternatively* determined by the observer O' *at rest* in the *moving train* using the relation

$$l_0 = v. \Delta t. \quad ..... \quad ..... \quad (2.2)$$

Now, let *another observer* O, standing on the ground at the *fixed point* F, also measures the *time required* by the "Entire length of the *running train* to move past the fixed point F, i.e. him". He uses his *clock* (which is *at rest* relative to him, i.e. relative to the ground), to register the times t_S^0 and t_R^0 as the end points S and R of the *running train* move past him.

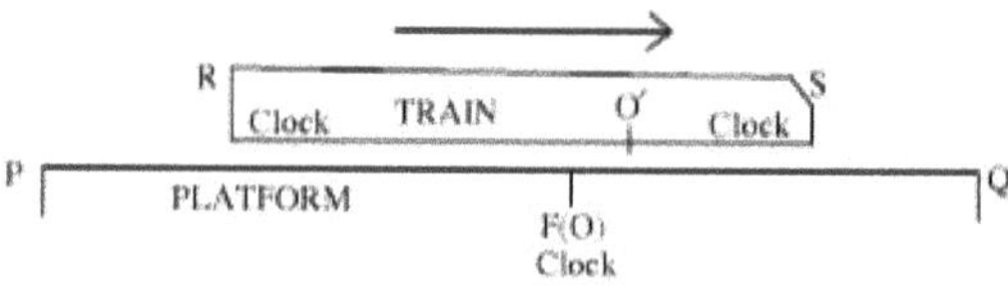

Fig . 6

Then
$$t_R^0 - t_S^0 = \Delta t_0 \text{ (say)}$$
is the *required time interval.* The length l of the *running train,* as determined by the observer O (standing on the ground) is then given by

$$l = v. \Delta t_0.. \quad ..... \quad ...... \quad (2.3)$$

We are, however, *not certain* whether the two measurement processes would provide the *same value* for l_0 and l, as given by the Eq.(2.2) and Eq.(2.3).

To resolve this question, we need to consider the intervals Δt_0 and Δt. The time interval Δt_0 between the *events* that "The ends S and R of the *running train* move past the fixed point F on the ground", is measured by a *single clock* at rest at the *fixed point* F, and the time interval Δt between the *same two* events is determined by *two clocks* at *two different positions* S and R. The time interval Δt_0 is then the *proper time interval* and Δt the *non-proper time interval* between the same two events. We therefore have, by Eq. (2.1a),

$$\Delta t_0 = \Delta t \sqrt{1 - v^2/c^2} \ . \qquad \ldots\ldots \quad \ldots\ldots \qquad (2.4)$$

Thus, the length of the *running train*, as determined by the observer O *at rest* on the ground (platform), would be given by

$$l = v. \Delta t_0 \qquad\qquad \text{[by Eq.(2.3)]}$$
$$= v. \Delta t \sqrt{1 - v^2/c^2} \qquad \text{[by Eq. (2.4)]}$$

and then, by using Eq. (2.2),

$$l = l(v) = l_0\sqrt{1 - v^2/c^2} \ . \qquad \ldots\ldots \quad \ldots\ldots \qquad (2.5)$$

Eq. (2.5) shows that "The *length* l of a train, as determined by an observer relative to whom the train is *running* with a velocity v, *gets shorter* by a factor $\sqrt{1 - v^2/c^2}$ than its *length* l_0, measured by an observer *at rest* relative to the train".

2.6.2 Moving Observer, Object at Rest.

We next turn to determine the *length* **L** of the platform PQ of the station as observed from the *running train*.

Let L_0 be the length of the platform PQ as measured by an *observer* O on the platform using the usual method, that is, by laying down a metre stick along the length PQ and counting the number (including fraction) of times the stick has to be laid to reach from the end P to the end Q.

The length L_0 can *alternatively* be determined by the *observer* O on the platform as follows (Fig.7).

The observer O measures the *time required* by "*Any fixed point,* say O'(an *observer* standing at O'), in the *running train* to traverse the entire length of the platform from P to Q". He determines this *time interval* by means of *two synchronized clocks* placed at the two ends P and Q of the platform. Let two persons standing at P and Q register the times, say t_P and t_Q, when the *fixed point* O' (observer at O') in the *running train* moves past the ends P and Q, respectively. Then

$$t_Q - t_P = \Delta t\,(\text{say})$$

is the *required time interval.* Let v be the uniform velocity of the running train. Then the length L_0 of the platform, as *alternatively* determined by the observer O on the platform, is given by

$$L_0 = v.\Delta t. \qquad \ldots\ldots \quad \ldots\ldots \qquad (2.6)$$

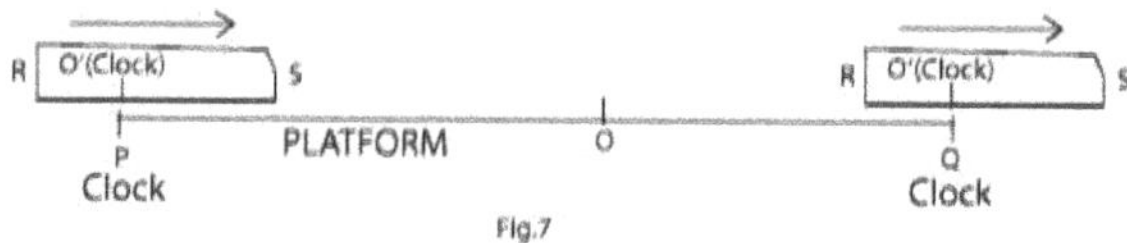

Fig.7

Now, let the observer at O' in the *running* train, register in his *clock* the times t_p^0 and t_q^0, when he just passes the end points P and Q of the platform. Let

$$t_q^0 - t_p^0 = \Delta t^0.$$

Then as measured by the observer O', moving together with the *running train* with velocity v relative to the platform, the length L of the platform is given by

$$L = v.\Delta t^0 \qquad \ldots\ldots \quad \ldots\ldots \qquad (2.7)$$

As in the previous case [Sec.2.6.1], here also we are *not sure* whether the two methods of measurement of the length of the platform would yield the *same value* for L_o and L, as given by the Eq.(2..6) and Eq.(2.7).

We then, as before, consider the time intervals Δt^0 and Δt. The time interval Δt^0 required for the *event* that "The observer O' sitting in the *running train* traverses the entire length PQ of the platform", is measured by a *single clock at rest* with respect to the observer O', and the time interval Δt for the *same event* is measured by *two different clocks* situated at two *different points* P and Q of the platform. The time interval Δt^0 is then the *proper time interval* and Δt the *non-proper time interval* for the same event. Using Eq. (2.1a), we then have

$$\Delta t^0 = \Delta t\sqrt{1 - v^2/c^2}. \qquad \ldots\ldots \qquad (2.8)$$

Thus, the length L of the platform as determined by the observer O' *at rest* in the *running train*, would, using the Eq.(2.7) and Eq.(2.8), be given by

$$L = v.\Delta t^0 = v.\Delta t\sqrt{1 - v^2/c^2}$$

and therefore, using Eq.(2.6), by

$$L = L(v) = L_0\sqrt{1 - v^2/c^2} . \quad ..... \quad (2.9)$$

Eq.(2.9) shows that the *length L of the platform*, as measured by an observer *moving* with velocity v relative to the platform, gets *shorter* by a factor $\sqrt{1 - v^2/c^2}$ than its length L_0, measured by an observer *at rest* on the platform.

2.6.3 Contraction in Length, Proper Length.

Considering together the results of the Sec.2.6.1 and Sec.2.6.2, we arrive at the following .

The length of a train *running* with a velocity v, gets *shortened,* i.e., *contracted,* by a factor $\sqrt{1 - v^2/c^2}$, when measured by an observer *at rest* on the platform(ground), and the length of the platform gets *contracted* by the same factor $\sqrt{1 - v^2/c^2}$, when measured by an observer *at rest* in the *running* train.

It is our common experience that, when a train speeds past a platform of a station, the platform appears, for an observer sitting in the train, to be moving in the opposite direction with the same speed as the train.

The length of an object, as measured by an observer *at rest* with respect to the object, is called the "*proper length*" of the object. A frame of reference in which an object under observation is *at rest,* is called the "*proper frame*" for the object.

Generalizing the results of the above "thought experiments" with a train running past a platform of a station, we obtain the following result:
"The length of an object as measured by an observer gets *contracted* from its *proper length* by the factor $\sqrt{1 - v^2/c^2}$

when there is a *relative motion* with velocity v between the observer and the object along its length."

In the Theory of Relativity this is known as the *Relativity of Length*, or more explicitly, as the "*Contraction in Length of an object moving along its length*".

Example — Let us, for instance, take the proper length of the train (the length measured by an observer at rest relative to the train) as $l_0 = 100$ m, and let the train be running with a velocity v (relative to the ground). Let the length of the running train as measured by an observer standing on the ground be $l = 99.5$m (according to the contraction in length of an object moving along its length). Then, since

$$l = l_0\sqrt{1 - v^2/c^2},$$

we have
$$99.5 = 100\sqrt{1 - v^2/c^2}$$

or
$$1 - v^2/c^2 = (0.995)^2 = 0.99$$

or
$$v^2/c^2 = 1 - 0.99 = 0.01$$

or
$$v/c = \sqrt{0.01} = 0.1$$

or
$$v = 0.1 \times c = 0.1 \times 3 \times 10^8 \text{ m/sec.}$$
$$= 30 \times 10^6 \text{ m/sec}$$
$$= 30{,}000 \text{ km/sec.}$$

Thus, for a *contraction in length* of 0.5m of a *running train*, as measured by an observer *stationary* on the ground, a 100m long (this length being measured by an observer *at rest* relative to the train) train has to run with a velocity
$$v = 30{,}000 \text{ km/sec.}$$
This enormously high value of the train's velocity shows why the contraction in length of moving objects are not perceived in our daily life; the velocities that we come across in ordinary life are much less than the velocity of light c.

———————

Chapter 3

The Lorentz Transformation and Some Consequences

Our objective is now to look for the relations of transformation, from one inertial reference system to another, that would keep the Maxwell Equations *invariant* (unchanged in form) in the two systems.

3.1 The Lorentz Transformation.

Let us consider two inertial systems of reference — rectangular Cartesian coordinate frames K(Oxyz) and K'(O'x'y'z') in uniform rectilinear motion relative to each other, together with two identical synchronized clocks placed at the respective origins O and O'. Following is the problem:

Given the space-time coordinates (x, y, z, t) of a *certain event* in the system K, to find the space-time coordinates (x', y', z', t') of the *same event* in the system K'. It is assumed, for simplicity, that during the relative motion between the coordinate frames K and K', the x and x'-axes coincide with each other, the axes of y and y' are parallel, and so are the axes of z and z', the frame K' moves with a uniform velocity v relative to the frame K along the positive direction of the common x-x'-axis. It is further assumed that the origins O and O' were coincident at the time $t = t' = 0$ as shown by the clocks at O and O' [Fig.1, Sec.1.4].

The transformation equations sought for are derived straightaway from the *two fundamental Postulates of Einstein* [Sec.1.11], and these equations are the following:

$$x' = \frac{x - vt}{\sqrt{1 - \frac{v^2}{c^2}}}, \quad y' = y, \quad z' = z, \quad t' = \frac{t - \frac{v}{c^2}x}{\sqrt{1 - v^2/c^2}} \quad \text{..... (3.1)}$$

This set of equations is known as the *Lorentz Transformation**. Solving the Eq. (3.1) for x, y, z, t, we obtain**

$$x = \frac{x' + vt'}{\sqrt{1 - v^2/c^2}}, \quad y = y', \quad z = z', \quad t = \frac{t' + \frac{v}{c^2}x'}{\sqrt{1 - v^2/c^2}}. \quad (3.2)$$

This is again a set of *Lorentz Transformation equations* with the exception of the relative velocity v being replaced by $-v$. Here the transformation gives the space-time coordinates (x, y, z, t) of *an event* in the frame K from the known values (x', y', z', t') of the space-time coordinates of the *same event* in the frame K', where the system K moves with velocity $-v$ relative to the system K', that is, along the negative direction of the common x-x'-axis.

If $v \ll c$, or if c gets *infinitely large*, then Eqs.(3.1) reduce to

$$x' = x - vt, \quad y' = y, \quad z' = z, \quad t' = t.$$

This is the set of our familiar Galilean transformation, Eqs.(1.1).

We now test whether under a Lorentz transformation the velocity of propagation of light retains the *same value c* in both the systems of reference K and K'.

Let a light signal (pulse of light) be transmitted from the origin O of the reference system K at time $t = 0$ along the positive x-axis. If c is the velocity of light in the system K, this light pulse proceeds according to the relation:

$$x = ct.$$

We are to derive the corresponding relation for the propagation of the light signal in the reference system K'. Substituting $c.t$ for x in the first and fourth of the Lorentz transformation equations, we obtain

$$x' = \frac{(c-v)t}{\sqrt{1-v^2/c^2}} \quad \text{and} \quad t' = \frac{t-(v/c)t}{\sqrt{1-v^2/c^2}} = \frac{(c-v)t}{c\sqrt{1-v^2/c^2}},$$

---

*The derivation of the Lorentz Transformation equations has been omitted here. A simple derivation of these equations is given in the Appendix.

**Solution of the Eqs.(3.1) for x, t, in terms of x', t', is also given in the Appendix.

and hence $$\frac{x'}{t'} = c, \ \text{or} \ x' = ct'.$$

Thus, the velocity of light in the system K' also turns out to be c. This result, however, is naturally expected, as the Lorentz transformation is essentially derived on the basis this concept, viz., "The Constancy of the velocity (speed) of light" — the Second Postulate of Einstein.

The Lorentz transformation *fails* for velocities $v > c$, and $v = c$. In fact, the *finite velocity* c of light in free space is the *limiting velocity of nature*, which can neither be surpassed, nor even attained, by any *material object*. Here, we keep the light (electromagnetic wave) beyond the realm of material objects, as light travels with velocity $v = c$.

3.2 Some Consequences of the Lorentz Transformation.

Let us now turn to some of the consequences of the Lorentz transformation. We consider two inertial reference frames K(x, y, z, t) and K' (x', y', z', t'), where K' moves with a uniform velocity v relative to K along the positive direction of the common x-x'-axis. Then the Lorentz transformation equations from the frame K to the frame K' are the following:

$$x' = \frac{x - vt}{\sqrt{1 - v^2/c^2}}, \ y' = y, \ z' = z, \ t' = \frac{t - (v/c^2)x}{\sqrt{1 - v^2/c^2}}. \ \ \ldots \ (3.1)$$

3.2.1 Relativity of Simultaneity.

Let two events (say, the flashes of two lights) take place at two different points (x_1, y, z) and (x_2, y, z) in the frame K at the *same time* $t_1 = t_2 = t$ (say). Then the corresponding times t_1' and t_2' in the frame K' are given by

$$t_1' = \frac{t - (v/c^2)x_1}{\sqrt{1 - v^2/c^2}} \ \text{and} \ t_2' = \frac{t - (v/c^2)x_2}{\sqrt{1 - v^2/c^2}} \ \ (\text{as} \ t_1 = t_2 = t).$$

Since $\quad x_1 \neq x_2$, it follows that $t_1' \neq t_2'$.

Thus, *two events* at different places occurring *simultaneously* for one observer, are *not simultaneous* for another observer, who is *moving* relative to the former. This is the *Relativity of Simultaneity.*

We have seen this earlier in the illustration of a *running train* approaching one light source and receding from another light source, in the Sec.2.4.

3.2.2 Relativity of Length: The Fitzgerald-Lorentz Contraction.

(a) Let the length of a straight stick (or rod) lying *at rest* in the system K' be l_o , as measured by an observer at rest in K'. The stick is placed along the common x-x'-axis in the system K', and let the coordinates of its end points be $(x_1',0,0)$ and $(x_2', 0, 0)$, with $x_1' < x_2'$, so that
$$l_o = x_2' - x_1' .$$
The question is: "What would be the length of this stick, when it is *moving relative to* and n measured by an observer (at rest) in the system K? "

To get the answer, we require to find the points x_1 and x_2 in the system K, where the beginning and the end of the stick lie on the x-axis, at any *single instant of time t* of the system K. Let,

$$x_2 - x_1 = l.$$

Then, from the first equation of the Lorentz transformation (from the system K to the system K'), viz.

$$x' = \frac{x - vt}{\sqrt{1 - v^2/c^2}}$$

we have
$$x - vt = x'\sqrt{1 - v^2/c^2} ,$$
and therefore,

$$x_1 - vt = x_1'\sqrt{1 - v^2/c^2}$$
and
$$x_2 - vt = x_2'\sqrt{1 - v^2/c^2} .$$

Hence,
$$x_2 - x_1 = (x_2' - x_1')\sqrt{1 - v^2/c^2} ,$$
or
$$l = l_o\sqrt{1 - v^2/c^2} . \qquad ..... \qquad ..... \qquad (3.3)$$

As the stick is *moving* with the velocity v relative to the system K, it therefore follows that:

"The length of a straight rigid stick, moving along its length with a velocity v, is *contracted* by the factor $\sqrt{1 - v^2/c^2}$."

Alternatively, the above result is obtained straightaway if we remember that the measurement of length of the stick is made at a *single instant of time* t in the system K. Thus, from the first equation of the Lorentz transformation, viz.,

$$x' = \frac{x - vt}{\sqrt{1 - v^2/c^2}},$$

we have $\qquad \Delta x' = \frac{\Delta x - v.\Delta t}{\sqrt{1 - v^2/c^2}} = \frac{\Delta x}{\sqrt{1 - v^2/c^2}},$ (since $\Delta t = 0$),

or $\qquad\qquad \Delta x = \Delta x'\sqrt{1 - v^2/c^2}.$

Hence, writing $\qquad \Delta x' = l_o$ and $\Delta x = l,$

we get $\qquad\qquad l = l_o\sqrt{1 - v^2/c^2}.$ $\qquad$ .... (3.3)

(b) Now, let the same stick be placed *at rest* in the system K along the common x-x'-axis, and let the coordinates of its end points be $(x_1, 0, 0)$ and $(x_2, 0, 0)$, with $x_1 < x_2$, so that we have

$$l_o = x_2 - x_1.$$

This time the question is: "What would be the length of this stick as measured by an observer in the system K', who, together with the system K', is *moving relative to the stick* which is placed *at rest* in the system K"?

Let x_1' and x_2' ($x_1' < x_2'$) be the points on the x-x'-axis in the system K', where the two ends of the stick lie at any *single instant of time t'* of the system K'. Let

$$x_2' - x_1' = l'.$$

Then, from the first equation of the Lorentz transformation, we have

$$x_1' = \frac{x_1 - vt_1}{\sqrt{1 - v^2/c^2}} \quad \text{and} \quad x_2' = \frac{x_2 - vt_2}{\sqrt{1 - v^2/c^2}},$$

so that $\qquad\qquad x_2' - x_1' = \frac{x_2 - x_1 - v(t_2 - t_1)}{\sqrt{1 - v^2/c^2}}.$

Also, from the fourth equation of the Lorentz transformation, viz.,

$$t' = \frac{t - (v/c^2)x}{\sqrt{1 - v^2/c^2}},$$

we have $\qquad t_1' = \frac{t_1 - (v/c^2)x_1}{\sqrt{1-v^2/c^2}}$ and $t_2' = \frac{t_2 - (v/c^2)x_2}{\sqrt{1-v^2/c^2}}$.

Since $\qquad\qquad\qquad t_1' = t_2' = t',$

we have $\qquad\qquad t_1 - (v/c^2)x_1 = t_2 - (v/c^2)x_2 ,$

so that $\qquad\qquad t_2 - t_1 = (v/c^2)(x_2 - x_1) .$

Hence, $\qquad x_2' - x_1' = \frac{(x_2-x_1) - v(v/c^2)(x_2-x_1)}{\sqrt{1-v^2/c^2}}$

$$= \frac{(x_2-x_1)(1-v^2/c^2)}{\sqrt{1-v^2/c^2}}$$

or $\qquad\qquad x_2' - x_1' = (x_2 - x_1)\sqrt{1 - v^2/c^2} ,$

and therefore, $\qquad l' = l_0\sqrt{1 - v^2/c^2} .$ $\qquad$ .... (3.4)

The results obtained in the above two Cases **(a)** and **(b)**, are in conformity with the First postulate of Einstein, viz., the *Principle of Relativity* — "Equivalence of any two inertial systems of reference."

We can now state the following:

"If v is the *relative velocity* between two inertial reference systems, then the length of a straight stick (or rod) lying at rest in one of them, along the direction of their *relative motion*, gets *contracted* by the factor $\sqrt{1 - v^2/c^2}$, when measured by an observer (at rest) in the other reference system."

This also has been seen earlier in the illustration of a *train running past a platform*, in the Sec.2.6.1 and Sec.2.6.2.

The Special Theory of Relativity, in fact the Lorentz transformation, thus accounts for the "Fitzgerald-Lorentz Hypothesis of Contraction of a *moving object* by the factor $\sqrt{1 - v^2/c^2}$ along its direction of motion.

3.2.3 Relativity of Time Interval: The Einstein Time Dilation.

Let us consider a clock placed at rest at a *fixed point* $(x'$, $0, 0)$ in the system K' (x', y', z', t'), emitting signals (ticks) at an interval
$$\Delta t' = t'_2 - t'_1 = 1 \text{ sec (say)}.$$
Measured in the system K (x, y, z, t), relative to which the system K' (together with the clock fixed in it), is moving with a velocity v, the time interval $\Delta t'$ is registered as Δt. Then, by the fourth equation of the Lorentz transformation from K' to K, we have

$$\Delta t = t_2 - t_1 = \frac{t'_2 - t'_1 + (v/c^2)(x'_2 - x'_1)}{\sqrt{1 - v^2/c^2}}$$

$$= \frac{t'_2 - t'_1}{\sqrt{1 - v^2/c^2}} \quad (\text{since } x'_2 = x'_1 = x'),$$

or $\qquad \Delta t = \dfrac{\Delta t'}{\sqrt{1 - v^2/c^2}} > \Delta t' = 1 \text{ sec.} \quad \dots\dots \quad (3.5)$

The interval of 1sec in the system K' appears to be elongated, or dilated, to the observer in the system K. As observed from K, the time that elapses between two ticks of the clock (at rest in K'), is not 1sec, but *greater than* 1sec.

Similarly, let the clock be placed *at rest* at a *fixed point* $(x$, $0, 0)$ in the system K. Then the interval
$$\Delta t = t_2 - t_1 = 1 \text{ sec}$$
between two signals (ticks) emitted by this clock, will be recorded by the observer in the system K', moving with velocity v relative to K, as:

$$\Delta t' = t'_2 - t'_1 = \frac{t_2 - t_1 - (v/c^2)(x_2 - x_1)}{\sqrt{1 - v^2/c^2}}$$

$$= \frac{t_2 - t_1}{\sqrt{1 - v^2/c^2}}, \quad (\text{since } x_2 = x_1 = x)$$

or $\qquad \Delta t' = \dfrac{t_2 - t_1}{\sqrt{1 - v^2/c^2}} > \Delta t = 1 \text{ sec.} \quad \dots\dots \quad (3.6)$

It then follows that: "A clock *moving* relative to an observer runs at a *slower rate* than when *at rest* relative to the observer".

This is the *Relativity of Time Interval* and is known as the *Einstein Time Dilation*. This has also been seen earlier in the "thought experiment" of a *running train* carrying a light source, a mirror and a clock, in the Sec.2.5.

3.2.4 Relativistic Velocity Addition Theorem (or Law).

We consider two inertial frames K and K', where K' is moving relative to K with a uniform velocity v along the positive direction of the common x-x'- axis. Let u' be the velocity of a particle P along the x'-axis with respect to the system K'. We want to determine the velocity u of the particle P with respect to the system K.

In the system K' the velocity u' is given by $u' = \dfrac{dx'}{dt'}$, and in the system K the velocity u by $u = \dfrac{dx}{dt}$.

By the Galilean transformation from K' to K, viz.,

$$x = x' + vt' \quad \text{and} \quad t = t',$$

we have

$$\frac{dx}{dt} = \frac{d}{dt'}(x' + vt')$$

$$= \frac{dx'}{dt'} + v\frac{dt'}{dt'}$$

or

$$u = u' + v, \qquad \text{.....} \quad \text{.....} \quad (3.7)$$

which is the well known Classical Velocity Addition theorem (or law). Now, using the first and fourth equations of the Lorentz transformation from K' to K, viz.,

$$x = (x' + vt')/\sqrt{1 - v^2/c^2}$$

and

$$t = (t' + vx'/c^2)/\sqrt{1 - v^2/c^2},$$

we obtain

$$u = \frac{dx}{dt} = \frac{dx}{dt'} \cdot \frac{dt'}{dt}$$

$$= \frac{\frac{dx'}{dt'} + v}{\sqrt{1 - v^2/c^2}}\frac{dt'}{dt} = \frac{u' + v}{\sqrt{1 - v^2/c^2}}\frac{dt'}{dt}$$

and

$$\frac{dt}{dt'} = \frac{1 + (v/c^2)\frac{dx'}{dt'}}{\sqrt{1 - v^2/c^2}} = \frac{1 + (v/c^2)u'}{\sqrt{1 - v^2/c^2}},$$

so that

$$u = \frac{u' + v}{\sqrt{1 - v^2/c^2}} \cdot \frac{\sqrt{1 - v^2/c^2}}{1 + (v/c^2)u'}$$

or

$$u = \frac{u' + v}{1 + u'v/c^2} . \qquad \text{.....} \quad \text{.....} \quad (3.8)$$

This is the *"Relativistic or Einstein's theorem (or law) of Addition of Velocities (both in the same direction)".*

From the Eq. (3.8), we can write

$$u' + v = u\,(1 + u'v/c^2)$$

or $\qquad u - v = u' - uu'v/c^2 = u'(1 - uv/c^2)$

or $\qquad u' = \dfrac{u - v}{1 - uv/c^2}\,.$ $\qquad$ .... $\qquad$ (3.9)

The Eq.(3.9) can also be obtained if we consider the Lorentz transformation from K to K', or if for the velocity v, we set $-v$ (the velocity of K with respect to K') in the Eq. (3.8).

3.2.5 Consequences of the Relativistic Velocity Addition Law.

The relativistic law of addition of velocities has some *revolutionary consequences.* For the Relativistic Addition of two velocities v_1 and v_2 in the same direction, we can write:

$$v = \frac{v_1 + v_2}{1 + v_1 v_2/c^2}$$

$$= \frac{c(v_1/c + v_2/c)}{1 + v_1 v_2/c^2}$$

$$= c.\frac{1 + v_1 v_2/c^2 - (1 + v_1 v_2/c^2 - v_1/c - v_2/c)}{1 + v_1 v_2/c^2}$$

or $\qquad v = c\left[1 - \dfrac{(1 - v_1/c)(1 - v_2/c)}{1 + v_1 v_2/c^2}\right]$ $\quad$ ..... $\quad$ (3.10a)

Eq.(3.10a) shows that, if $v_1 < c$ and $v_2 < c$, then $v < c$. The result holds for more than two velocities (all in the same direction) added successively. If either v_1 or v_2, or both, are *equal to c,* *only then* we get $v = c$, for

$$v = \frac{c + v_2}{1 + c\,v_2/c^2} = \frac{c + v_2}{1 + v_2/c} = c, \qquad ..... \qquad (3.10b)$$

and $\qquad v = \dfrac{c + c}{1 + c.c/c^2} = \dfrac{2c}{2} = c.$ $\qquad$ .... $\qquad$ (3.10c)

The Relativistic Velocity Addition law thus leads to the *incredible equation:*

$$c + c = c, \qquad ..... \qquad ..... \qquad (3.10d)$$

where c is the velocity of light in free space.

When we consider the *subtraction* of one velocity from another, that is, the addition of two velocities in the opposite directions, we have

$$v = \frac{v_1 - v_2}{1 - v_1 v_2 / c^2} \cdot \qquad ..... \qquad (3.11a)$$

If $v_1 = c$, we have, from the Eq.(3.11a)

$$v = \frac{c - v_2}{1 - c\, v_2 / c^2} = \frac{c - v_2}{1 - v_2 / c} = c. \qquad ..... \qquad (3.11b)$$

Thus, if any velocity $v\ (<c)$ is *added to, or subtracted from*, the *velocity of light c*, then the *resultant velocity* obtained is the *same velocity c.* This result, although very striking, should not be surprising to us, because, initially we had started with the hypothesis that: *"The velocity of light is the same c in every inertial system"*—the Second postulate of Einstein. The following important result therefore follows.

"By successive addition of velocities, each of which is less than or equal to the velocity c of light in empty space, and each of which is in the same direction, *it is not possible to obtain a velocity greater than c".*

The velocity c of light in free space, thus, plays the role of a limiting velocity.

3.3 Theory of Relativity and Experiments.

Among the different Ether and Emission Theories, no single theory could explain the results of all the related light propagation experiments carried out up to the early 20th century. The Special Theory of Relativity of Einstein, on the other hand, was amazingly successful in accounting for the results of not only all those experiments, but also the results of relevant experiments performed later. Further, the predictions of the

Special Theory of Relativity were all experimentally verified in future. We now consider two such experiments in the following.

3.4 Fizeau's Experiment and Relativistic Velocity Addition Law.

It was mentioned in the Sec.1.7.4(b) that, as predicted by Fresnel (in 1817) and later experimentally verified by Fizeau (in 1851 and 1853), the *velocity c'* of light in a moving medium is given, relative to the observer in the laboratory, where the medium (water) is moving (through a tube), by the formula:

$$c' = c_\omega \pm v_\omega \left(1 - \frac{1}{n^2}\right). \qquad ..... \qquad ..... \qquad (3.12)$$

Here, the velocity c' is measured relative to the observer *at rest* in the laboratory where the medium (water) is made to flow through a tube, $c_\omega = (c/n)$ is the velocity of light in motionless still water, c is the velocity of light in empty space (in the "stationary ether", as presumed at that time), n the refractive index of water and v_ω the velocity of water relative to the tube carrying the flowing water. The positive or negative sign occurs according as the light and water move in the same or opposite directions. According to Classical velocity addition law, the *velocity c'* should be given by

$$c' = c_\omega \pm v_\omega .$$

The Relativistic velocity addition law of Einstein, on the other hand, gives the *velocity c'* as:

$$c' = \frac{c_\omega \pm v_\omega}{1 \pm c_\omega . v_\omega / c^2} = (c_\omega \pm v_\omega)\left(1 \pm c_\omega . \frac{v_\omega}{c^2}\right)^{-1}. \quad ... \quad (3.13)$$

Since, $c_\omega/c = 1/n = 1/1.33 < 1$, and since the non-relativistic velocity v_ω of the flowing water is much less than c (as $v_\omega \approx 7$ m/sec and $c = 3 \times 10^8$ m/sec), we can neglect $(v_\omega/c)^2$, and therefore, we have

$$c' = (c_\omega \pm v_\omega)(1 \mp c_\omega v_\omega / c^2)$$

$$= c_\omega \pm v_\omega \mp (c_\omega/c)^2 . v_\omega \mp c_\omega . (v_\omega/c)^2$$

$$= c/n \pm v_\omega \mp (1/n)^2 . v_\omega \mp (1/n). v_\omega^2/c$$

$$= c/n \pm v_\omega (1 - 1/n^2) \mp 0.000,000,123$$

and hence $$c' \approx \frac{c}{n} \pm v_\omega \left(1 - \frac{1}{n^2}\right). \qquad ..... \qquad (3.14)$$

The Fizeau experiment was repeated by Michelson and Morley in 1886, and later by Zeeman and others after 1914, confirming Fizeau's result with much greater precision.

The *"Relativistic Velocity Addition theorem"* of Einstein thus agrees very well with Fizeau's result (to within 1%). No hypothesis of "Partial-drag of light by Flowing water" is, therefore, necessary to account for the result of the Fizeau experiment. The result follows directly from *Einstein's Velocity Addition theorem.*

3.5 The Meson Experiments and the Relativity of Time Interval.

Unstable sub-atomic elementary particles can be treated as *fundamental clocks* of nature. Each type of such particles, on creation, disintegrates (breaks up) *at rest* into other particles, after a definite *"lifetime"* or *"mean-life"*. These particles, because of their property of instability, can be used as *clocks*, either *at rest* or *in motion.*

Light elementary sub-nuclear particles — *mesons (mu-meson or muon* and *pi-meson or pion*), with masses intermediate between those of an electron and a proton, originate in nature in the upper layers of the earth's atmosphere. High energy extra-terrestrial radiation from far outer space, known as *Cosmic Rays* (Primary Cosmic Radiation), impinges on the earth's atmosphere and interacts with the nuclei of the molecules constituting the atmosphere. The interactions generate another radiation (Secondary Radiation) consisting of mesons; these mesons subsequently disintegrate into other particles.

The *mu-mesons* (discovered in 1936) are produced at a height of around 5 to15 km above the sea level, and a large fraction of them reaches the earth's surface with velocity v close to that of light c ($v > 0.99c$). They are the most abundant type of mesons found in cosmic rays at the earth's surface. The *pi-mesons* (discovered in1947) are generated much higher in the upper atmosphere; they then move toward the ground with velocity comparable to that of light c and spontaneously decay (break up) during flight, producing *mu-mesons.*

The mesons can also be created in the laboratory by means of nuclear splitting. The *average lifetime* of an artificially produced *mu-meson*, measured *at rest* in the laboratory, comes about to be
$$\Delta t_o = 2.2 \times 10^{-6} \text{ sec.} = 2.2 \text{ microsec.} \quad \text{.... } (3.15)$$
and the *average lifetime* of an artificially produced *pi-meson,* measured *at rest,* is
$$\Delta t_o = 2.5 \times 10^{-8} \text{ sec.} \quad \text{..... ..... } (3.16)$$
These time intervals are therefore the *proper lifetimes* of the two kinds of mesons.

3.5.1 The mu-meson Experiment and Relativity of Time.

According to Classical mechanics, mu-mesons during their *mean lifetime* of 2.2 microsec, could on the average, travel *no more* than a distance of
$$l = c . \Delta t_0$$
$$= 3 \times 10^8 \text{ m/sec} \times 2.2 \times 10^{-6} \text{ sec} \qquad \text{[by Eq.(3.15)]}$$
$$= 660 \text{ m only,}$$
and no mu-mesons would then have been observed at the ground level. The *Relativistic time dilation* gives a mu-meson, moving toward the earth's surface with a velocity say, $v = 0.995c$ (as $v > 0.99c$), the *average lifetime (non-proper),* measured by an observer on the ground, as:
$$\Delta t = \frac{\Delta t_o}{\sqrt{1-v^2/c^2}} = \Delta t_o / \sqrt{1-(0.995)^2}$$

$$= \Delta t_o / \sqrt{1-0.99} = \Delta t_o / \sqrt{0.01} = \Delta t_o / 0.1$$

$$= 10 \times 2.2 \times 10^{-6} = 22 \times 10^{-6} \text{ sec.} \quad \text{[by Eq.(3.15)]}$$
During this time interval, the distance travelled by this mu-meson, as measured by the observer on the ground, is given by
$$L = v . \Delta t = 0.995c . \Delta t$$

$$= 0.995 \times 3 \times 10^8 \text{m/sec} \times 22 \times 10^{-6} \text{sec}$$

$$= 6567 \text{ m} > 6.5 \text{ km.}$$

It is thus the *Relativity of Time interval* that accounts for the mu-meson showers observed on the earth's surface.

3.5.2 The pi-meson Experiment and Relativity of Time.

If there were *no Relativity of time intervals* in nature, pi-mesons during their *mean lifetime* of 2.5×10^{-8} sec, could, on the average, travel *no more* than a distance of

$$l = c.\Delta t_o$$

$$= 3 \times 10^{8} \text{ m/sec} \times 2.5 \times 10^{-8} \text{sec} \qquad [\text{by Eq.(3.16)}]$$

$$= 7.5 \text{ m.}$$

Laboratory experiment was performed with pi-mesons at Columbia University. They were generated by splitting the nucleus of atoms and a beam of pi-mesons ejected with velocity v close to that of light c ($v \approx 0.99c$).

The "half-life" period of pi-mesons, *brought to rest,* was determined. This half-life is *defined* to be "The *time interval* in which *half* of a number of pi-mesons, existing at any instant, disintegrates". The half-life period of pi-mesons *at rest* was measured to be

$$\Delta t_o = 1.77 \times 10^{-8} \text{sec} \quad ..... \quad ..... \quad (3.17)$$

and this time period is the *proper time interval* for the pi-meson.

Now, according to Classical mechanics, half of the pi-mesons in a beam could, on the average, travel *at most* a distance of

$$d_1 = c.\Delta t_o$$
$$= 3 \times 10^{8} \text{ m/sec} \times 1.77 \times 10^{-8} \text{ sec} \quad [\text{ by Eq. (3.17)}]$$
$$= 5.3 \text{ m only.}$$

However, it was experimentally found that, in a beam of pi-mesons moving with velocity $v = 0.99c$, the intensity of the beam reduced to *half* of its *initial value* at a distance of nearly 39 m from the source. The above two results appear to be contradictory. What then is the explanation?

The average *half-life* period Δt of the *moving* pi-mesons in the aforesaid beam (as measured in the laboratory frame of reference and which, therefore, is a *non-proper* time interval), is given, according to the *Relativity of time interval*, by

$$\Delta t = \Delta t_o/\sqrt{1 - v^2/c^2} = \Delta t_o/\sqrt{1 - (0.99)^2}$$

$$= \Delta t_o/\sqrt{1 - 0.98} = \Delta t_o/\sqrt{0.02} \ \text{sec}$$

$$= 1.77 \times 10^{-8}/0.14 \qquad \text{[by Eq. (3.17)]}$$
$$= 12.6 \times 10^{-8} = 0.126 \times 10^{-6}$$

or $$\qquad\qquad\qquad \Delta t \approx 0.13 \times 10^{-6} \text{sec}. \qquad ..... \quad ..... \qquad (3.18)$$

During this time interval, the average distance travelled by a pi-meson in the beam, as measured in the laboratory frame of reference, is given by

$$d_2 = 0.99c.\,\Delta t$$
$$= 0.99 \times 3 \times 10^8 \times 0.13 \times 10^{-6} \text{ m} \quad [\text{ by Eq. (3.18)}]$$
$$= 0.386 \times 10^2 \text{ m} = 38.6 \text{ m} \approx 39 \text{ m}.$$

Thus, it is the *Relativity of time interval* that accounts for the distance of nearly 39 m measured in the laboratory, travelled by the pi-mesons during their *half-lives.*

Each of the above two experiments, respectively, with mu-mesons and pi-mesons, therefore, establishes the *reality* of the *Einstein Time Dilation,* or the *Relativity of Time Interval.*

3.6 Experiments and the Fitzgerald-Lorentz Contraction.

The null results of the Michelson-Morley experiments were accounted for by the *contraction in length* of the *two arms* of the optical interferometer apparatus used in the experiments.

Lorentz attempted to explain the "Contraction of an object moving with a velocity v through the stationary ether (presumed to exist at that time) by the factor $\sqrt{1 - v^2/c^2}$ along its direction of motion" by means of his *"Theory of electrons"* applied inside the material objects.

However, in the Special Theory of Relativity *no such explanation* is necessary. The "Fitzgerald-Lorentz contraction" is derived straightaway (without assuming the existence of any 'ether' medium) from the Lorentz transformation [Sec.3.2.2 (a) & (b)], which in turn is obtained from the two Postulates of the Special Theory of Relativity.

The "Contraction of a moving object along its direction of motion," follows as a *"Natural consequence of the process of measurement made by the observer, whenever there exists a relative motion between the object and the observer."* No "medium of ether", nor any "electronic phenomena within the object", is needed for this contraction.

It may be interesting to note that the results of the *Meson Experiments*, as given in the Sec.3.5.1 and Sec.3.5.2, can also be accounted for by means of the "Contraction in Length of an object moving along its length", as given in the following sections.

3.6.1 The mu-meson Experiment and Contraction of Length.

Let us consider a mu-meson (or muon) produced in nature at an altitude of, say 6.5 km, above the ground and moving toward the earth with a velocity $v = 0.995c$. The moving muon 'sees'(an 'observer' sitting on it sees) the vertical distance $L_0 = 6.5$ km in the atmosphere to be moving past 'him' in the opposite direction with the same velocity v, and therefore, 'sees' the distance L_0 *contracted* by the factor $\sqrt{1 - v^2/c^2}$.

The 'observer', positioned at rest on the muon, then measures (with respect to the frame of reference fixed with the muon) the distance travelled by him, as

$$L = L_0 \sqrt{1 - v^2/c^2} = L_0\sqrt{1 - (0.995)^2}$$
$$= 6.5 \times \sqrt{0.01} = 6.5 \times 0.1 \text{ km}$$
$$= 0.65 \text{ km} = 650 \text{ m}.$$

The *time (proper)* required by this muon, in its own frame of reference, to traverse this distance is given by

$$\Delta t_o = L/v = 650 \text{ m}/(0.995 \times 3 \times 10^8 \text{ m/sec})$$
$$= 650/(2.985 \times 10^8) \text{ sec}$$
$$= 2.18 \times 10^{-6} \text{ sec}$$
$$< 2.2 \times 10^{-6} \text{ sec} = 2.2 \text{ microsec.}$$

Now, it is known that 2.2 microsec. is the *average lifetime* (proper time) of a muon.

Thus, it is because of the *contraction in length of the vertical distance to be traversed,* that a muon, originated at a height of 6.5 km above the sea level, makes it possible to reach the ground with a velocity $v = 0.995c$ within its average lifetime of 2.2 microsec, and this accounts for the abundance of mu-mesons detected on the earth's surface.

3.6.2 The pi-meson Experiment and Contraction of Length.

Let us consider a pi-meson artificially created in the laboratory and moving with a velocity $v = 0.99c$. The moving pion 'sees' (an 'observer' sitting on it sees) the laboratory distance $L_0 = 38$ m (say) to be moving in the opposite direction with the same velocity v, and therefore, 'sees' the distance L_0 *contracted* by the factor $\sqrt{1 - v^2/c^2}$. The 'observer' placed on the pion measures the distance (in the frame of reference fixed with the pion) travelled by him, as

$$L = L_0\sqrt{1 - v^2/c^2} = L_0\sqrt{1 - (0.99)^2} = L_0\sqrt{0.02}$$
$$= 38 \times 0.14 \text{ km} = 5.3 \text{ m}$$

The *time (proper)* required by this pion, in its own frame of reference, to travel this distance is given by

$$\Delta t_o = L/v = 5.3 \text{ m}/(0.99 \times 3 \times 10^8 \text{ m/sec})$$
$$= 5.3/(2.97 \times 10^8) \text{ sec} = 1.78 \times 10^{-8} \text{ sec,}$$

which is the *half-life* of the pion in the frame in which it is at rest.

Thus, it is the *contraction in length of the distance to be traversed,* that enables a pion to travel a distance of 38 m or more (in the laboratory frame of reference), which is much greater than the distance of 5.3 m as travelled by the pion in its own (rest) frame of reference.

Hence, each of the two experiments with mu and pi-mesons, respectively, also proves the "Fitzgerald-Lorentz contraction of a *moving object* (here the distance traversed) along its direction of motion relative to the observer".

Chapter 4

Relativistic Mechanics

4.1 Bridging the Classical and Relativistic Mechanics.

Evolution in physical science does not discard the previously known and established laws, but only puts up boundaries to their sphere of validity. When new theories come up, their consequences have to be in conformity with the relevant older theories in the domain in which the former theories have already been proven to be correct.

Classical Newtonian mechanics holds for velocities v which are much smaller compared to the velocity of light c, for velocities v less than 1000 km/sec, that is, for $v < 0.003\,c$. The Special Theory of Relativity is valid for velocities v near the speed of light c ($v \approx 0.99\,c$). Such velocities are usually observed for ions, electrons and other elementary particles.

It has been seen that the Lorentz transformation of the Special Theory of Relativity reduces to the Galilean transformation of Classical mechanics for velocities $v <\!< c$, and also that the Relativistic velocity Addition rule for velocities v_1 and v_2 , viz.,

$$v = \frac{v_1 + v_2}{1 + v_1 v_2 / c^2}$$

reduces to the Classical velocity Addition rule

$$v = v_1 + v_2,$$

when the velocities v_1, v_2 are $\ll c$. The laws of Classical Newtonian mechanics remain the *same in form,* we say, remain *Invariant,* or rather, *Covariant* under a Galilean transformation, but *not* under a Lorentz transformation.

To reconcile the Classical and Relativistic mechanics we require to modify the Classical laws in such a way that the new

Relativistic laws may reduce to the older Classical laws in case of velocities $v << c$, and also that they remain *covariant* under a Lorentz transformation.

In Classical mechanics, from Newton's Second Law of Motion — "The rate of change of momentum of a particle is proportional to the impressed force":

$$\vec{F} = \frac{d\vec{p}}{dt} = \frac{d}{dt}(m\vec{v}), \quad ..... \quad ..... \quad (4.1)$$

there follows the *Law of Conservation of Momentum*—"In the absence of any external forces the momentum of a moving particle (or object) is conserved":

$$\vec{p} = m\vec{v} = \text{constant, when } \vec{F} = 0. \quad ...... \quad (4.2)$$

4.2 Variation of Mass (Inertial).

Let us consider the mutual collision between two identical, perfectly elastic and smooth spheres. It can be shown that (we omit the mathematical details) the Law of Conservation of Momentum—"The sum of the momenta of two particles (spheres) before collision is equal to the sum of the momenta of the particles after collision"—would preserve the same form in Classical as well as in Relativistic mechanics, provided the *mass (inertial)* of a particle (or an object) moving with a velocity $\vec{v}$, is *defined* by the relation :

$$m = m(v) = \frac{m_o}{\sqrt{1-v^2/c^2}}, \quad ...... \quad (4.3)$$

where m_o is our familiar *classical mass* and $m = m(v)$ is called the "*relativistic mass*" of the particle or object (sphere in this case).

The classical mass m_o is called the *rest mass* of the particle (or sphere), as this is the value which $m(v)$ assumes when $v = 0$, that is, when the particle is *at rest.* The rest mass m_o is also called the *proper mass,* for it is the mass of an object as measured, like the *proper time* [Sec.2.5.1] and *proper length* [Sec. 2.6.3], in an inertial frame in which the particle (object) is *at rest.*

The *Relativistic momentum* then assumes the form

$$\vec{P} = m\vec{v} = \frac{m_o\vec{v}}{\sqrt{1-v^2/c^2}} \quad \ldots\ldots \text{ (4.4)}$$

which reduces to the Classical form of momentum

$$\vec{p} = m_o\vec{v}$$

for values of $v \ll c$. It can be shown (we omit the calculations) that the key factor $1/\sqrt{1-v^2/c^2}$ in the Eq. (4.4) arises because of the Relativity of Time measurements.

As in Classical mechanics, if also in Relativistic mechanics a Force is measured by the "Time rate of change of momentum", then Newton's Second Law of motion:

$$\vec{F} = \frac{d\vec{p}}{dt} = \frac{d}{dt}(m\vec{v})$$

assumes the following *modified form*:

$$\vec{F} = \frac{d\vec{P}}{dt} = \frac{d}{dt}\left(\frac{m_o\vec{v}}{\sqrt{1-v^2/c^2}}\right). \quad \ldots\ldots \quad \text{(4.5)}$$

One of the most striking results of the Relativistic mechanics is that, "The inertial mass of a particle is not an invariant property of the particle, but it increases with its velocity (speed)". It is seen from Eq.(4.3) that:

$$m(v) \rightarrow m_o \text{ as } v \rightarrow 0, \text{ and } m(v) \rightarrow \infty \text{ as } v \rightarrow c.$$

Such an increase in the inertial mass with velocity (speed) was first experimentally observed in the motion of radioactive beta rays (electrons) by Kaufmann in 1902 [Sec.1.9].

4.3 Equivalence of Mass (Matter) and Energy.

In Classical mechanics, the kinetic energy (energy due to motion), say T_C, of a particle is measured by the work done by an externally impressed force in raising the speed of the particle from zero to some positive value. Then

$$T_c = \int_{v=0}^{v} \vec{F}.\vec{ds}$$

where $\vec{F}.\vec{ds}$ is the work done by the force $\vec{F}$ in displacing the particle through a distance $\vec{ds}$. Taking, for simplicity, the motion along the x-axis only, we have

$$T_c = \int_{v=0}^{v} F\,dx = \int_0^v \frac{d}{dt}(m_o v)dx = \int_0^v m_o \frac{dv}{dt}dx = m_o \int_0^v \frac{dx}{dt}dv$$

$$= m_o \int_0^v v\,dv$$

or $$T_c = m_o[v^2/2]_0^v = \frac{1}{2}m_o v^2 , \quad \text{......} \quad (4.6)$$

where m_o is the *Classical mass* of the particle.

In *Relativistic mechanics*, the Kinetic Energy, say T_R, of a particle, is measured by using the *relativistic definition* of force, as given by the relativistic equation of motion, that is, the *modified form of the Newtonian equation of motion*, viz., Eq. (4.5). We then have

$$T_R = \int_{v=0}^{v} F\,dx = \int_0^v \frac{d}{dt}(mv)dx = \int_0^v d(mv)\frac{dx}{dt} = \int_0^v (mdv + vdm)v$$

or $$T_R = \int_{v=0}^{v}(mvdv + v^2 dm). \quad \text{.....} \quad (4.7)$$

Now, since the *relativistic mass* $m = m(v)$ is given by

$$m = m_o/\sqrt{1 - v^2/c^2} , \quad \text{.......} \quad (4.3)$$

we have $$m^2(1 - v^2/c^2) = m_0^2 ,$$

or $$m^2 c^2 - m^2 v^2 = m_0^2\, c^2,$$

from which, taking differentials of both sides, we obtain the following:

$$2c^2 m\,dm - 2m^2 v\,dv - 2v^2 m\,dm = 0$$

or $$c^2 dm = mv\,dv + v^2 dm \quad \text{.....} \quad (4.8)$$

Using the Eqs.(4.7) & (4.8), we then have

$$T_R = \int_{v=0}^{v} c^2 dm = c^2 \int_{m=m_o}^{m} dm = c^2(m - m_o)$$

or
$$T_R = mc^2 - m_o c^2. \qquad \text{.....} \quad (4.9)$$

The Eq.(4.9), on writing the value of m, given by Eq.(4.3), leads to the following:

$$T_R = \frac{m_o c^2}{\sqrt{1-v^2/c^2}} - m_o c^2 \qquad \text{.....} \quad (4.10)$$

$$= m_o c^2 \left[(1 - v^2/c^2)^{-\frac{1}{2}} - 1 \right]$$

$$= m_o c^2 \left[\left(1 + \frac{1}{2}\frac{v^2}{c^2} + \frac{3}{8}\frac{v^4}{c^4} + \text{... ...} \right) - 1 \right]$$

or
$$T_R = \frac{1}{2} m_o v^2 + \frac{3}{8} m_o \frac{v^4}{c^2} + \cdots \qquad \text{....} \quad (4.11)$$

The Eq.(4.11) then, gives the *Relativistic form* of the Kinetic Energy of a particle.

When $\frac{v}{c} \ll 1$, neglecting $\frac{v^2}{c^2}$, we obtain, from Eq.(4.11),
$$T_R = \frac{1}{2} m_o v^2 = T_c \quad [\text{ by Eq. (4.6)}] \quad \text{....} \quad (4.12)$$

It is thus seen that, the *Relativistic expression* T_R for the Kinetic Energy, given by the Eq.(4.11), reduces in the limit
$$\frac{v}{c} \to 0,$$
to the corresponding *Classical expression* T_c.

Very significantly, if in the Eq.(4.9), we set $mc^2 = E$, and call E the *Total Energy* of the particle, then we have
$$E = m_o c^2 + T_R, \qquad \text{......} \quad \text{.....} \quad (4.13)$$

or, by the Eq.(4.10), $\qquad E = \frac{m_o c^2}{\sqrt{1-v^2/c^2}}, \qquad \text{... ..} \quad \text{.....} \quad (4.14a)$

and by the Eq.(4.11),

$$E = m_o c^2 + \frac{1}{2} m_o v^2 + \frac{3}{8} m_o \frac{v^4}{c^2} + \cdots \qquad \ldots \ (4.14b)$$

The quantity $m_o c^2$ is called the *Rest Energy* of the particle, where m_o has already being termed as the *rest mass* of the particle [Sec.4.2]. The *rest energy* is the *energy* of the particle when $v = 0$, that is, when the particle is *at rest*, and that is, when its kinetic energy is *zero*. The *rest energy* is denoted by E_o, so that we obtain

$$E_o = m_o c^2. \qquad \ldots \ldots \qquad \ldots \ldots \quad (4.15)$$

The rest energy $E_o = m_o c^2$ of a particle is an *invariant quantity* for all velocities and it has *no significance* for the motion of the particle.

The *most revolutionary result* of the Special Theory of Relativity is the famous equation of Einstein:

$$E = m\,c^2, \qquad \ldots \ldots \qquad \ldots \ldots \quad (4.16)$$

where E is the *total energy* (= *rest energy* + *relativistic kinetic energy*), m is the *relativistic mass* and c the velocity of light in empty space.

Thus, it is seen that *Energy* and *Mass* are *equivalent;* they are *not essentially* two different physical entities, as was supposed in the pre-relativistic days, but the former is c^2 times the latter. Although the derivation of the Eq.(4.16) in Einstein's original work was concerned only with kinetic energy, experiments performed later proved, as predicted by Einstein, that the relation is valid in general, for all kinds of energy.

Prior to the advent of the relativity theory, the "Law of Conservation of Mass" and the "Law of Conservation of Energy" were regarded as two distinct "Laws of Nature". But now, with the equation $E = m\,c^2$, it becomes clear that each of the two laws has no separate identity. The two are *merged* into a *single law* — the *"Law of Conservation of a certain physical entity:*

Mass-Energy". This entity appears sometimes as mass and sometimes as energy.

4.4 Experimental Verification of Mass Energy Equivalence.

It is now known that, "Energy possesses mass", or equivalently, "Mass contains energy", and also that "Energy can be transformed into mass (matter) and mass (matter) into energy". The equation $E = m c^2$ has been verified in numerous experiments.

4.4.1 Energy turned into Mass.

The *Relativistic kinetic energy* of a particle may *exceed* its *rest energy*. A very high energy (velocity approaching that of light) proton colliding with another proton *at rest*, is capable of producing new particles. In such a collision, a *proton-antiproton pair* is *created* in addition to the two initially colliding protons. The *rest energy* necessary to produce the two new particles is provided by the *energy* of the *incident proton*. The *kinetic energy* of the incident proton is *partly* transformed into the *rest masses* of the products. This is an example where energy is converted into mass (matter).

4.4.2 Mass turned into Energy.

There are examples of phenomena in which matter (mass) is turned into energy. For instance, the *annihilation* of a proton with an antiproton, or that of an electron with a positron (antielectron), where, in each case, the entire *rest masses of the pair* of particles are transformed into the *energy of electromagnetic radiation.*

Atomic (Nuclear) Reactors have been constructed in which suitable material is converted into energy by means of controlled *nuclear disintegration*, and this energy is utilized, through some complex technical process, to generate electricity.

If *one gram* of matter is transformed into energy, the amount of *energy liberated*, according to the equation $E = m c^2$, is given in the following.

$$E = 1.c^2 = (3 \times 10^{10})^2$$
$$= 9 \times 10^{20} = 0.9 \times 10^{21} \text{ ergs}$$
$$= 0.9 \times 10^{14} \text{ Joules} \quad (10^7 \text{ergs} = 1 \text{ Joule})$$
$$= 90 \times 10^{12} \text{ joules} = 90 \text{ trillion joules.}$$

Expressed in terms of *heat energy* this is

$$E = 0.238 \times 90 \times 10^{12} \text{ calories}$$
$$= 21 \times 10^{12} \text{ cal} = 21 \text{ trillion cal.}$$

Among the illustrations, where *mass is being converted into energy,* the most dreadful one is that of the Atomic (Nuclear) Bomb. The atom bomb that destroyed Hiroshima in 1945 lost only about 1gm of its initial fissionable material (uranium), and being transformed into energy, this small amount of matter produced an energy of nearly 10^{14} Joules ($\approx 21 \times 10^{12}$ calories), which is equal to the energy liberated in the explosion of more than 20,000 tons (20 kilotons) of TNT.

The energy of very large explosions is usually expressed in terms of the energy released in the explosion of *one ton* of the explosive chemical TNT (trinitrotoluene). The unit *1ton TNT* is defined as the energy of 1 billion (= 10^9) calories, or 4.2 billion Joules.

Note: -

1 calorie = 4.184 joules $\approx$ 4.2 joules,
1 joule = $1/4.2 \approx$ 0.238 cal,
1 ton TNT = 10^9(=1billion or 1giga) cal = 4.2 billion joules,
1 kiloton (1,000 ton) TNT = 10^{12}(=1trillion or 1tera) cal,
20 kiloton TNT = 20 trillion cal = $20 \times 10^{12} \times 4.2$ joules
$$=0.84 \times 10^{14} \text{joules} \approx 0.9 \times 10^{14} \text{ joules}$$
$$= 90 \text{ trillion joules.}$$

Chapter 5

Four-Dimensional Space-Time Continuum

The entire universe is a "Four-Dimensional Space-Time Continuum". What does this mean?

5.1 Time as the Fourth Dimension.

It is known that "Space is a Three-Dimensional Continuum". This means that, it is possible to specify the position of a point, say P, in *Space* by three numbers x, y, z, which represent the distances of the point P measured from a given point(origin) O in space, in the respective directions of three rectangular Cartesian coordinate axes OX, OY, OZ, and also that, there exists an infinite number of points around the point *(x, y, z)* whose positions can be denoted by *(x+dx, y+dy, z+dz)*, where *dx, dy, dz* may be chosen as small as we like. Because of these two properties, Space is said to be *Three-Dimensional* and a *Continuum*, the axes OX, OY, OZ indicating the *three dimensions* and the numbers x, y, z being known as the *three coordinates* of the point P.

Any **event** in the universe takes place at a *point (x, y, z)* of the three-dimensional space and at some point (moment or instant) of *time t. Without the time t an event has no existence.* In addition to the three space dimensions, *time,* therefore, needs to be regarded as another dimension — the *Fourth Dimension.*

5.2 The Four-Dimensional Minkowski 'World'.

The aggregate of *all events* in the entire universe, then, constitutes a *Four-Dimensional Continuum* — the *Space-Time Continuum*, which was first termed *'World'* by Minkowski in 1907. Any *event* in this World is specified by four numbers x, y, z, t, and is called a *"World Point"*, being denoted by *(x, y, z, t).*

As in the case of the three-dimensional space continuum, here also, an indefinite number of world points (*x+dx, y+dy, z+dz, t+dt*), where *dx, dy, dz, dt* are arbitrarily small, exists in the neighbourhood of the world point *(x, y, z, t)*.

Before the advent of the Theory of Relativity, *time* was regarded as an absolute continuum, independent of all events and the system of reference. In Classical mechanics, this concept is tacitly manifested in the fourth equation of the Galilean transformation:

$$t' = t .$$

In Relativistic mechanics, however, *time* is not independent and is *related with the space coordinates*, as expressed in the fourth equation of the Lorentz transformation:

$$t' = \frac{t - (v/c^2)x}{\sqrt{1 - v^2/c^2}}$$

Hence, instead of treating space and time separately, as in Classical mechanics, they are treated together and in the same manner in Relativistic mechanics.

Minkowski, in 1908, suggested that "Space by itself and time by itself are to fade away (into mere shadows), and only a kind of union of the two, the 'World' in itself, would preserve an independent reality." We can then write the *time t* in terms of the *distance*, say *w*, as

$$w = c\,t, \qquad \text{..... (5.1)}$$

where *c* is the speed of light in empty space, a constant in the entire 'World'. This *w* is then regarded as the *fourth coordinate, the time coordinate, x, y, z,* being the first three, the space coordinates, of a *"Four-dimensional orthogonal (rectangular) system of coordinates".*

Now, from the first and fourth of the Lorentz transformation equations, we have

$$x'^2 - c^2 t'^2 = \frac{(x - vt)^2}{1 - v^2/c^2} - c^2 \frac{(t - vx/c^2)^2}{1 - v^2/c^2}$$

$$= \frac{x^2 + v^2 t^2 - 2vxt - c^2 t^2 - v^2 x^2/c^2 + 2vxt}{1 - v^2/c^2}$$

$$= \frac{x^2(1 - v^2/c^2) - c^2 t^2 (1 - v^2/c^2)}{1 - v^2/c^2} = x^2 - c^2 t^2$$

so that
$$x'^2 - c^2 t'^2 = x^2 - c^2 t^2. \qquad \text{.......} \qquad (5.2)$$

As $y' = y$, $z' = z$, it then follows that
$$x'^2 + y'^2 + z'^2 - c^2 t'^2 = x^2 + y^2 + z^2 - c^2 t^2. \quad \text{...} \quad (5.3)$$

The Lorentz transformation equations, therefore, transform the expression
$$x^2 + y^2 + z^2 - c^2 t^2 \quad \text{into} \quad x'^2 + y'^2 + z'^2 - c^2 t'^2$$
and vice-versa. We can, henceforward, characterize the Lorentz transformation as a set of equations under which the expression

$$x^2 + y^2 + z^2 - c^2 t^2 \quad \text{i.e.,} \quad x^2 + y^2 + z^2 - w^2 \quad \text{... (5.4)}$$

remains invariant.

The Lorentz transformation can be characterized, more simply if, instead of the *real time* coordinate $w \, (= c \, t)$, we introduce the *imaginary time* coordinate

$$i \, w = i \, ct, \quad \text{where } i = \sqrt{-1},$$

as the *fourth coordinate* of the Minkowski World. On writing
$$x_1 = x, \; x_2 = y, \; x_3 = z, \; x_4 = iw \, (= ict), \quad \text{where } i = \sqrt{-1},$$
for the four coordinates of the reference system K, and similarly with dashes for the coordinates of the system K', the Eq.(5.3) can be written as:
$$x_1'^2 + x_2'^2 + x_3'^2 + x_4'^2 = x_1^2 + x_2^2 + x_3^2 + x_4^2. \quad \text{..... (5.5)}$$

The General Lorentz transformation in the *"Four-dimensional Space-Time* World*"* is then defined as a linear homogeneous transformation of the four coordinates $x_1, x_2, x_3, x_4 (= ict)$, which keeps the quadratic form:
$$x_1^2 + x_2^2 + x_3^2 + x_4^2 . \qquad \text{.....} \quad \text{.....} \quad (5.6)$$
an *invariant.*

The Four-dimensional Minkowski World possesses a close similarity with our familiar Three-dimensional Space, which is Euclidean — the Space where Euclidean Geometry holds. The

Minkowski World can be regarded as a Four-dimensional *Euclidean Space* with *imaginary time coordinate*.

Finally, it may be remarked that, while the Four-dimensional Space-Time Continuum of the Special Theory of Relativity, i.e., the Minkowski World, is a Four-dimensional Euclidean Continuum, the Space-Time Continuum of the General Theory of Relativity is a *Non-Euclidean Continuum*. Any discussions about this, however, are beyond the scope of the present book.

Appendix

Simple Derivation of the Lorentz Transformation

Given the space-time coordinates *(x,y,z,t)* of a *certain event* in the inertial system K(O x y z t), the problem is to find the space-time coordinates *(x',y',z',t')* of the *same event* in another inertial system K'(O'x'y'z't').

For simplicity, it is assumed that during the relative motion between the frames K and K' the x and x'-axes remain coincident, the axes of y, y' remain parallel and so also the axes of z, z', and the system K' moves with a uniform velocity v with respect to the system K along the positive direction of the common x-x'-axis. It is also assumed that the origins O and O' of the two frames were coincident at the time $t = t' = 0$ as shown by two identical clocks, synchronized with each other, and placed at the respective origins.
Let us consider a plane light wave front proceeding along the positive x-x'-axis in the system K. The observer in the system K describes it by the equation

$$x = c\,t$$

or
$$x - c\,t = 0. \quad\quad \text{.....} \quad\quad \text{.....} \quad\quad \text{(A1)}$$

In the system K', according to the *Second Postulate* of relativity, this light wave propagates with the *same velocity c,* and therefore, the observer in K' describes it by the similar relation

$$x' - ct' = 0. \quad\quad \text{.....} \quad\quad \text{......} \quad\quad \text{(A2)}$$

The space-time points (events) which satisfy the Eq.(A1) must also satisfy the Eq (A2). This means that the relation

$$x' - ct' = a\,(x - ct), \qquad \ldots\ldots \quad \ldots\ldots \quad (A3)$$

where a is a constant (at most a function of the velocity v only), is satisfied in general; for, according to the Eq.(A3), the vanishing of $(x - ct)$ implies, and is implied by, the vanishing of $(x' - ct')$. Similarly, considering a plane light wave front proceeding along the negative x-x'-axis, we have

$$x' + ct' = b\,(x + ct). \qquad \ldots\ldots \quad \ldots\ldots \quad (A4)$$

Solving the Eqs (A3) & (A4) we get

$$x' = \frac{b+a}{2}x + \frac{b-a}{2}.ct$$

$$ct' = \frac{b-a}{2}.x + \frac{b+a}{2}.ct$$

Writing $A = \frac{b+a}{2}$ and $B = \frac{b-a}{2}$, these equations can be written as

$$x' = Ax + Bct \qquad \ldots\ldots \quad \ldots\ldots \quad (A5a)$$

$$ct' = Bx + Act \qquad \ldots\ldots \quad \ldots\ldots \quad (A5b)$$

Now since, at $t = 0$, $x = 0$, by hypothesis we have $t' = 0$, $x' = 0$, and since the origin O' of the frame K' moves with a uniform velocity v relative to the frame K, the point $x' = 0$ must be identical with the point $x = v\,t$. Therefore, from Eq.(5a) we get

$$0 = A.\,vt + B.\,ct$$

or

$$B = -A.\frac{v}{c},$$

so that from Eq.(A5a) $\qquad x' = A(x - vt) \quad \ldots\ldots \quad \ldots\ldots \quad (A6a)$

and from Eq.(A5b) $\qquad ct' = -A\frac{v}{c}x + Act,$

or $\qquad\qquad t' = A\left(t - \frac{v}{c^2}x\right). \qquad \ldots\ldots \quad (A6b)$

Thus, the problem now reduces to that of evaluation of the constant A. To find the value of A, we turn to the *First Postulate* of relativity.

We place a measuring stick (a metre-stick) *at rest* along the x'-axis of the system K' such that the coordinates of its end points are

$$x' = 0 \text{ and } x' = 1 \quad (y' = 0, \ z' = 0).$$

Measured from the system K, these points have at the *time t = 0*, according to the Eq.(A6a), the coordinates

$$x = 0 \text{ and } x = \frac{1}{A} \quad (y = 0, \ z = 0).$$

Hence, the *length* of the metre-stick, as measured by the observer in the system K at the *time t = 0*, is

$$1/A. \quad \text{........} \quad \text{.....} \quad \text{.......} \quad \text{.....} \quad (A7)$$

Next, we place the same metre-stick *at rest* along the x-axis of the system K such that the coordinates of its end points are

$$x = 0 \text{ and } x = 1, \text{ with } (y = 0, z = 0).$$

Measured from the system K', these points have at the *time t'=0*, the coordinates which are determined below.

At the *time t' = 0*, we have the following:

(i) at $x = 0$, by Eq.(6b) we have $t = 0$, and therefore, from Eq.(A6a) we get $x' = 0$.

(ii) at $x = 1$, by Eq.(A6b) we have $0 = A(t - v/c^2)$, or $t = v/c^2$, and therefore, from Eq.(A6a) we get $x' = A(1 - v^2/c^2)$.

Thus, the required coordinates are

$$x' = 0 \text{ and } x' = A(1 - v^2/c^2), \text{ with } (y' = 0, \ z' = 0).$$

Hence, the *length* of the metre-stick, as measured by the observer in the system K' at the *time t' = 0*, is

$$A(1 - v^2/c^2). \quad \text{.....} \quad \text{.....} \quad \text{.......} \quad (A8)$$

By the *First Postulate* the two measurements by the two observers in the systems K and K', as given by (A7) and (A8), must be *equal*, so that

$$\frac{1}{A} = A(1 - v^2/c^2)$$

and hence
$$A = \pm\, 1/\sqrt{1 - v^2/c^2}\,.$$

Now it is known that $v = 0$ should lead to the Identity Transformation: $x' = x$, $t' = t$. Therefore, the positive sign must be chosen for A, and so, we have

$$A = 1/\sqrt{1 - v^2/c^2}\,. \qquad \cdots \quad \cdots \qquad (A9)$$

Hence, from the Eqs. (A6a) & (A6b) we obtain

$$\left.\begin{aligned}
x' &= \frac{x - vt}{\sqrt{1 - v^2/c^2}} \\[1em]
y' &= y, \\[1em]
z' &= z \\[1em]
t' &= \frac{t - (v/c^2)x}{\sqrt{1 - v^2/c^2}}
\end{aligned}\right\} \qquad ..(A10a)$$

This is the set of celebrated **Lorentz Transformation equations.**

If we solve for x, t in terms of x', t' from the Eqs.(A10a), we get*

$$\left.\begin{aligned}
x &= \frac{x' + vt'}{\sqrt{1 - v^2/c^2}} \\[1em]
y &= y' \\[1em]
z &= z' \\[1em]
t &= \frac{t' + (v/c^2)x'}{\sqrt{1 - v^2/c^2}}
\end{aligned}\right\} \qquad .. (A10b)$$

This is again a set of Lorentz Transformation equations with relative velocity $-v$, i.e., where the system K is moving with velocity $-v$ relative to the system K', i.e., moving with a velocity v along the negative direction of the common x-x'-axis.

---

* Solution:

Since
$$x' = (x - vt)/\sqrt{1 - v^2/c^2}$$
and
$$t' = (t - vx/c^2)/\sqrt{1 - v^2/c^2},$$
we have

$$x' + vt' = \frac{x - vt + vt - (v^2/c^2)x}{\sqrt{1 - v^2/c^2}} = \frac{x(1 - v^2/c^2)}{\sqrt{1 - v^2/c^2}}$$

$$= x\sqrt{1 - v^2/c^2},$$

and

$$t' + vx'/c^2 = \frac{t - vx/c^2 + vx/c^2 - (v^2/c^2)t}{\sqrt{1 - v^2/c^2}} = \frac{t(1 - v^2/c^2)}{\sqrt{1 - v^2/c^2}}$$

$$= t\sqrt{1 - v^2/c^2}.$$

Hence,

$$x = (x' + vt')/\sqrt{1 - v^2/c^2} \quad \& \quad t = (t' + vx'/c^2)/\sqrt{1 - v^2/c^2}.$$

Index

** ** ** **